The Political Economy of Hydropower Dependant Nations

"Imad Ahmed's deep dive into the story of Zambia's development of and dependency on hydropower provides a timely input into how countries with hydro resources should think about their energy transitions to zero-net emissions systems that provide reliable, affordable, and clean energy for all. Like many other countries, Zambia relies on an aging hydropower stock, developed when the speed and the full force of climate impacts seemed either unlikely or far into the future. For countries like Zambia, how can hydro play an optimized role in a diversified energy system, efficiently, serving the environment, people, and industry, as regional weather patterns change? How will they be financed with levels of indebtedness climbing past prudent limits in Zambia and across the region? What is hydropower's role in the energy systems of the future? This book is an excellent place to look back to move forward."

—Rachel Kyte, Dean, *The Fletcher School, Tufts University, USA; former UN SG Special Representative for Sustainable Energy; former World Bank Group Vice President and Special Envoy for Climate Change*

"Imad Ahmed has written a fascinating and meticulously researched book on Zambia, which details the various challenges posed by hydroelectric power. This book is a must-read for all interested in Zambia and energy policy in the developing world."

—Pradeep Chhibber, *Professor of Political Science, University of California, Berkeley, USA*

"One of the many devastating effects of climate change is that it exposes a fifth of humanity to energy insecurity. Through this brilliant analysis of the country of Zambia and the challenges ahead for its hydropower sector, Imad Ahmed shines a bright light on this emerging crisis and what must be done about it. This book is a must-read for anyone interested in getting beneath the headlines of the effects of climate shocks and taking meaningful action."

—Bhaskar Chakravorti, *Dean of Global Business, The Fletcher School, Tufts University, USA*

"Which comes first in developing economies, economic growth or environmental sustainability? Imad Ahmed's deep dive into Zambian hydropower successfully pulls together many diverse and important themes. This book provides a refreshing look at where traditional responses to hydropower can have adverse effects. Ahmed's conclusions might help developing countries 'leapfrog' towards environmental sustainability in emerging markets while providing the power for growth and standards of living that their citizens deserve."

—Michael Mainelli, Professor, *Alderman & Sheriff of the City of London, Executive Chairman, Z/Yen Group, UK*

Imaduddin Ahmed

The Political Economy of Hydropower Dependant Nations

A Case Study of Zambia

palgrave
macmillan

Imaduddin Ahmed
The Bartlett Faculty of the Built Environment
University College London
London, UK

ISBN 978-3-030-71265-5 ISBN 978-3-030-71266-2 (eBook)
https://doi.org/10.1007/978-3-030-71266-2

This Palgrave Macmillan imprint is published by the registered company Springer Nature Switzerland AG.
The registered company address is: Gewerbestrasse 11, 6330 Cham, Switzerland

This book is dedicated to the intended beneficiaries of my PhD research: the people of Zambia. I hope this is helpful.

Foreword

The increasing urgency of the global climate crisis has rekindled enthusiasm for renewable energy, which has in turn encouraged greater scrutiny of a given country's energy mix. In western countries, this has generally taken the form of national commitments to decarbonisation of the energy supply, phasing out fossil fuels in favour of greener alternatives such as on-grid wind turbines and off-grid solar photovoltaics. By contrast in developing countries, there has been a long-standing preference for hydropower projects. Many policymakers, public officials and even the public themselves believe that hydropower is consistent with deep decarbonisation, and that hydropower projects are thus bankable on both ESG and economic grounds. The reality is rather more complex, as Dr. Ahmed's pioneering study of Zambia's dependence on hydropower reveals.

In the course of examining the political economy of Zambia's energy mix, Dr. Ahmed asks hard questions about how far hydropower is sustainable if extreme weather events, such as droughts caused by El Niño, reduce reservoir levels by as much as 95% and force firms to rely on diesel backup generators, which are sources of both carbon emissions and air pollution. Moreover, the proliferation of smaller-scale hydropower projects represents a significant source of embedded carbon; if brought on-stream, these dams are further destined to destroy natural carbon sinks. Hydropower undoubtedly has a significant role to play, but over-reliance comes at a cost and the externalities caused by these projects deserve closer attention.

When I agreed to supervise Dr. Ahmed's doctoral dissertation at the University College London's (UCL) Bartlett School of Sustainable Construction, I would not have been able to predict that one of the chief contributions of his work would be to climate change economics. In 2016, we agreed a revised scope of work with the International Growth Centre for a Department for International Development (DFID)-funded project titled, 'Electricity Crises in Zambia: Estimating the Costs of Unreliability at Firm Level,' which aimed to understand the consequences of nationwide power outages caused by droughts the previous year. The inability of the hydropower-dependent utility, ZESCO, to cope with these droughts, especially in the face of rising electricity demand, had, in turn, actively undermined Zambia's industrial policy, which had been to diversify growth away from the primary sector (mining and agriculture) with a policy mix that encouraged export-oriented, electricity-dependent manufacturing firms. Our task was to quantify the adverse welfare effects, from the perspective of these firms, of this unmet demand for electricity and to understand how far the policy response, which was to waive import duties on diesel backup generators, represented an adequate and effective mitigation strategy. Our method was to survey a stratified sample of manufacturing firms, geographically located in Lusaka and the Copperbelt.

This research work, ably project-managed by Dr. Ahmed and field-managed by Mr. Graham Sianjase, rapidly revealed nuanced mitigation strategies employed by the wide range of firms sampled, which made it possible to assess wider environmental impacts. Dr. Ahmed's path-breaking research highlights the importance of the contributions made by modern environmental economics to infrastructure economics as a discipline. Moreover, in exploring winners and losers, at the firm level, he was able to focus attention on the political economy of energy policy, and by doing so to formulate cogent policy recommendations for regulators, legislators and managers of energy utilities. One can only hope that other researchers will follow in his footsteps as the climate crisis accelerates and forces hard choices about reconciling sustainability and growth agendas.

London, UK D'Maris Coffman

Acknowledgements

Thank you, Professor D'Maris Coffman and Dr. Priti Parikh, my primary and second supervisors. I was very lucky to have had your guidance, and am lucky that our interests intersect so well. Thank you, D'Maris, for introducing me to the historical institutional framework of path dependence, and suggesting the importance of the Tennessee Valley Authority. Thank you for opening multiple opportunities, including funding this PhD journey. Thank you, Priti and Dr. Ian Hamilton, my co-second supervisors, for testing the durability of my findings and making sure that I was meeting my milestones. Thank you, Professor Michelle Baddeley, for bringing me onto the Zambia project and into the PhD. Thank you, Professor Jim Meikle, for guiding me on this project at its early stages. Thank you, Dr. Musiliu Oseni, for outlining the initial framework of the manufacturing research in Zambia. Thank you to my viva examiners Professors Roberto Zoboli and Tim Broyd for your incisive feedback that has helped improve the presentation of this research. Thanks to my colleagues at the University College London (UCL) for supporting me: Dr. Bernard Tembo; Dr. EY Song and Vivien Kizilcec, who gave me crash courses in software; Helen Pascoe; Yukiko Fujimoto; Michael Franklyn and Anita Treso. Thanks to UCL and Goodenough College for funding my PhD, housing and living expenses.

Thank you, Graham Sianjase, my ever-cooperative and professional research field manager, without whom this project would have been a whole lot more difficult and a lot less successful. Your turn next. Thank you, Beauty Nkosha, Johanna Mwila, Mundia Kayamba, Nandi Ngwenye and Tapiwa Christine Janda, for your many hours of hard work in collecting the primary data from the manufacturing firms. Thank you for your time and patience to all those who participated as anonymous respondents in the surveys. Thank you, Chiti Mataka, Puseletso Mwakalombe, Mukuka Mubanga, Temwani Chirwa and Kachinga Wankunda Phiri of Zambia Electricity Supply Corporation (ZESCO) Ltd. Thank you to Alfred Mwila and Cletus Sikwanda of the Energy Regulation Board. Thank you, Winford Simwanza, Levison Phiri and Allan Chivunda of the Department of Energy. Thank you, Mwila Daka and Sikabele Chikuba of the Ministry of National Development Planning. Thank you, Chipego Zulu of the Zambia Association of Manufacturers. Thank you, Edson Nkonde of the Zambia Meteorological Department. Thank you, Goodson Sinyenga, Sheila Shimwambwa-Mudenda, Joseph Tembo and Litia Simbangala of the Central Statistical Office of Zambia. Thank you, Anthony Bwembya of the Patents and Companies Registration Agency. Thank you, Sombo Kaweza and Sampa Chilanga of the Zambia Development Agency. Thank you, Jacob Lungu, Peter Zgambo, Chungu Kapembwa and Wilson Phiri, Chisala of the Bank of Zambia. Thank you, Mikko Marttala of KPA Unicon. Thank you, Ana Hajduka and Cathy Oxby of Africa GreenCo. Thank you, Yimbilanji Sichone (researcher), Wamulume Kalabo (former Chairman of Zambia Chamber of Commerce and Industry, ZACCI), Kimberly Lyon (World Bank), Michael Mainelli, Bruce Bouchard, Jim Friedlander, Chibelushi Maxwell Musongole, Natty Chilundiki, Sunday Mikoka, Sachin Gathani, Dimitri Stoelinga, Gareth Walsh and Kim Lyon.

Thank you, Shivani Haria, with whom I co-authored a blog post; Dr. Dennis Chiwele, Kunda-Chibale Chima, Miljan Sladoje and Herryman Moono of the International Growth Centre (IGC) Zambia; as well as the rest of the editorial economics team, including Anand Rajaram, Jamie Green, Alar Cartan and Matei Alexianu, who with Shivani provided critical feedback and suggestions on a 59-page working paper and blog. That work and feedback has been incorporated here. Thanks to the IGC for the funding to the UCL Bartlett School of Sustainable Construction that enabled the primary data collection for this research.

Thank you to all my teachers, mentors and coaches. Thank you, Baba, Amme, Nano, Zar Aslam, Dadi, Professor Pradeep Chhibber, Farasat Bokhari, Nigar Ahmad and Tariq Siddiqi, Jugnu Mohsin, Hajrah Mumtaz, Amin Ahmed, Wajahat Bokhari, Kristen Zecchi, Professor Laurent Jacque, Dorothy Orszulak, Thomas Schmidheiny, Zain Latif, Shilesh Muralidhara, Sandra Rwamushaija, Kampeta Sayinzoga for opening new opportunities and broadening my horizons. Thank you for holding me professionally accountable, Issam, and intellectually accountable, Sheherbano, and for editing this work for the better. Thank you, Maaria Siddiqi, for the patience and support in so many ways throughout this endeavour that could only be asked of a loving spouse.

Contents

About the Book

This book aims to inform better energy policy in hydropower-dependent countries which are vulnerable to climate shocks. Using Zambia as a case study, it examines the causes and impact of increasing energy insecurity as global warming affects a fifth of the world population that live in hydropower-dependent countries facing drought.

Chapter 2 explains the underlying economic theory and evidence connecting energy and economic development through the medium of industrialisation. Chapter 3 reviews the literature on the impact of power outages on manufacturing. Chapter 4 assesses the impact of the 2015 and 2016 power outages on Zambia's manufacturing. Chapter 5 uses the historical institutional framework of path dependence to explain why Zambia's system of energy provision did not prevent the 2015 and 2016 power outages. Chapter 7 recommends how to address Zambia's power crisis. Chapter 8 relays the broad lessons for achieving low-carbon energy security in drought-affected countries.

Elements of this book have been previously been published:

1. Ahmed, I., Baddeley, M., Coffman, D. M., Meikle, J., & Sianjase, G. (2019). *The cost of power outages to Zambia's manufacturing firms*. London. doi: F-41408-ZMB-1
2. Ahmed, I. (2020a). Explaining Rwanda's prioritisation of rural electrification over rural clean drinking water through institutional path dependency. *Structural Change and Economic Dynamics, 54*(September), 186–201. https://doi.org/10.1016/j.strueco.2020.05.001

3. Ahmed, I., Parikh, P., Sianjase, G., Coffman, D. (2020). The impact decades long dependence on hydropower in El Niño impact-prone Zambia is having on carbon emissions through diesel backup generation. *Environmental Research Letters, 15*(December). https://doi.org/10.1088/1748-9326/abb6a1
4. Haria, S., & Ahmed, I. (2020). *No Increasing tariffs to prevent another electricity crisis in Zambia*, *International Growth Centre blog*. https://www.theigc.org/blog/increasing-tariffs-to-prevent-another-electricity-crisis-in-zambia

About the Author

Imaduddin Ahmed holds a PhD in Infrastructure Economics and Finance from the University College London, where he is a research fellow. Previously as a Transactions Advisor at the Rwanda Development Board, he helped the Rwandan energy utility execute thermal and renewable power purchase agreements.

Book Abstract

Zambia is one of at least 32 developing countries with a combined population of 1.4 billion people that are more than 25% dependent on hydropower for their grid electricity generation and which have experienced El Niño droughts. El Niño-induced droughts are projected to increase in frequency as average global temperatures rise. Droughts in 2015, 2016 and 2019 rendered low hydropower reservoir levels in Zambia, constraining electricity generation capacity and resulting in unprecedented power outages. Notwithstanding these frailties, the national power utility ZESCO and system of grid electricity provision has subordinated the need to achieve system sustainability and diversify its energy supply assets to the path-dependent pursuits of subsidising existing consumers and building more hydropower generation assets. The path-dependent approach was locked-in by earlier macro-level investment decisions, over which the World Bank had significant influence. Like 17 other countries, Zambia received World Bank investments for hydropower in the 1950s–70s to support mining or industrialisation. Path dependence ultimately undermined the Zambian government and ZESCO's attempts to enhance Zambians' welfare. Through analysis of primary collected data, this research shows that the consequences of ZESCO's path dependence, which resulted in power outages, have had an adverse impact on the driver of Zambia's future industrialisation. The research shows that ZESCO's 'least regret' path going forward in terms of baseload power, climate impact, human health and financial lifecycle costs involves charging cost-recovery tariffs and diversifying its portfolio of generation assets. As global average temperatures and the frequency of droughts rise, these findings have implications for the formulation of a low-carbon energy policy for a fifth of the world's population living in hydropower-dependent countries affected by increasing incidence of drought.

Institutional Acronyms

CEC	Copperbelt Energy Corporation
CSO	Central Statistics Office
DfID	UK Department for International Development
ERB	Energy Regulation Board
IBRD	International Bank for Reconstruction and Development
IDA	International Development Association
IGC	International Growth Centre
IMF	International Monetary Fund
IPCC	Intergovernmental Panel on Climate Change
MDNP	Ministry of National Development Planning
MFEZ	Multi-facility economic zone
OECD	Organisation for Economic Cooperation
UCL	University College London
ZACCI	Zambia Chamber of Commerce and Industry
ZAM	Zambia Association of Manufacturers
ZCCM	Zambia Consolidated Copper Mines
ZDA	Zambia Development Agency
ZESCO	State-owned power utility, formerly known as Zambia Electricity Supply Corporation
ZICTA	Zambia Information and Communications Technology Authority

List of Figures

List of Tables

List of Boxes

CHAPTER 1

Introduction

This research aims to inform better energy policy in hydropower-dependent countries which are vulnerable to climate shocks. As global average temperatures rise, so too do the frequency and intensity of El Niño Southern Oscillation-induced droughts (Wang et al., 2017), which in turn threaten the reliability of hydropower. This research therefore focuses on the impact of energy insecurity arising from climate-induced drought, which, according to climate modelling, would also result in reduced water availability and crop yields (Conway et al., 2015).

As Box 1.1 illustrates, the exposure of developing countries to hydropower is material. Appendix A further illustrates that at least 32 countries that were not high-income with a combined population of 1.4 billion people were more than 25% dependent on hydropower and had experienced El Niño droughts.[1] Eighteen of these received early World Bank investments for hydropower predominantly to support industrialisation or mining (Appendix A). In Sub-Saharan Africa, 160 million grid electricity consumers lived in countries where hydropower accounted for more than half of total power supply (Falchetta et al., 2019). In 2015 and 2016, El Niño droughts, caused by warming of the Pacific Ocean, resulted in severe power crises in Kenya, Tanzania, Ghana, Zimbabwe and Zambia, with power outages, power rationing and switching to costlier backup

[1] The World Bank data did not give information for all countries such as Uganda and the Southern African Power Pool countries Lesotho and Swaziland.

I. Ahmed, *The Political Economy of Hydropower Dependant Nations*,
https://doi.org/10.1007/978-3-030-71266-2_1

diesel generators (Falchetta et al., 2019). Droughts also increased competition for water use between power generation, irrigation and municipal water supply (Falchetta et al., 2019). Hydrological crop and recursive dynamic computable general equilibrium analysis suggested that the climate variable could pull 2% of the Zambian population below the poverty line (Thurlow et al., 2012).

Box 1.1 The exposure of developing countries to hydropower for their grid electricity
As of 2015 (World Bank, 2020a)

- >58 countries >25% reliance on hydropower for domestic electricity production
  - 78% were not high-income
  - >35 countries, 74% of which were not high-income >50% reliant on hydropower
    - >32 countries not high-income countries >25% hydropower reliant experienced El Niño droughts (Appendix A)
      18 of these countries received early World Bank investments for hydropower (Appendix A)
    - >12 countries, 11 of which were not high-income >80% reliant on hydropower

Genre of countries:

- Heavily Indebted Poor Countries: 46% reliant on hydropower for electricity production (World Bank, 2020a)
- Least Developed Countries: 37% reliant on hydropower (World Bank, 2020a)
- Latin America and Caribbean: 45% grid electricity from hydropower (World Bank, 2020a)
- Sub-Saharan Africa: 160M grid electricity consumers live in countries >50% reliant on hydropower (Falchetta et al., 2019)
  - Almost entirely dependent on hydropower: DRC, Lesotho, Malawi and Zambia (Conway et al., 2015)

This research takes Zambia, a landlocked southern African nation of 17.3 million people (World Bank, 2019), as its case study. It is one of the at least 32 countries that was not high-income and was more than 25% reliant on hydropower, and one of the at least 18 of these countries that received early World Bank investments for hydropower predominantly to support industrialisation or mining. At the beginning of 2015, Zambia was almost entirely dependent on hydropower and was classed as a lower middle-income country (World Bank, 2019) with 33% grid electricity connection (ZICTA and CSO, 2018). Pre-independence energy investment was driven by colonial extraction of copper (Bayliss & Pollen, 2019, p. 7). Post-independence energy investment until 2015 had been dominated by the World Bank, which, guided by unbalanced growth theory, perpetuated energy generation for the purpose of copper extraction.[2] Themes relating to Zambia's history of energy investment as well as the climate resilience of its energy infrastructure are therefore applicable to a number of other hydropower-dependent developing countries.

The research was borne of an International Growth Centre (IGC) commission won by the University College London's (UCL) Bartlett School of Sustainable Construction and which commenced in October 2016. The IGC is largely funded by the UK's Department for International Development (DfID) (International Growth Centre, 2018). The commission sought an assessment of the impact of power outages on Zambia's manufacturing sector. The International Growth Centre aims to promote sustainable growth in developing countries. The original focus was on the impact of power outages on the manufacturing sector given the established associations between infrastructure on the growth of manufacturing and the growth of manufacturing on economic development. Africa in general has experienced a period of de-industrialisation as a result of poor policy (the pursuit of structural adjustments at the behest of the IMF and World Bank) rather than natural economic evolution (Stiglitz, 2017). Given Zambia's manufacturing resurgence since the 2000s, undersupply of energy due to climate-vulnerable power generation infrastructure poses a new threat to Zambia's industrialisation and more generally to the country's trajectory of economic development and diversification away from copper mining, whose export as a raw material dominates Zambia's exports (World Bank, 2020d). The mining sector's profits are largely

[2] Independence was achieved on 24 October 1964.

expatriated to foreign multinationals.[3] By contrast, this research's survey of large manufacturing firms found about half of respondents said that their company owners were Zambian.

Alongside other factors, recent El Niño droughts resulted in reduced power supply to the manufacturing sector and consequently reduced growth. Due to almost complete dependence on hydropower and low rainfall in 2015 and 2016 resulting in low reservoir levels (Mwila et al., 2017, p. v), Zambia's power utility ZESCO was forced to enact nationwide load shedding lasting upwards of 8 hours per day (Energy Regulation Board, 2017, p. 15). The power deficit reached 1000 MW in 2015 (41% of national installed capacity)[4] and reduced to 526 MW in 2016 (Energy Regulation Board, 2017, p. 15). Had ZESCO not mitigated its generation shortfall with the import of 785.2 GWh in 2015 (up from just 12.8 GWh in 2014) and 2184.9 GWh in 2016 (Energy Regulation Board, 2017, p. 9), the load shedding would have been even worse. Table 1.1 is a compilation of grid electricity consumption by sector collected from Energy Regulation Board reports. It shows that from 2015 to 2017 manufacturing's share of Zambia's electricity consumption fell from 4.6% to 4.1%, and that absolute consumption fell from 2015 to 2016, and had not recovered to its 2015 level by 2017. Figure 1.1 charts the growth rate of Zambia's manufacturing value-added in constant local currency figures from the World Bank's indicators. It shows that growth in the sector fell from 6.5% in 2014 to 1.9% in 2016.

Why should we focus on manufacturing, other than for the fact that this research was funded to focus on it, when its energy supply is a tenth of mining's and its contribution to GDP lags not only that of mining but of retail and construction (CSO, 2018)? First, manufacturing provides better value in terms of contribution to GDP and employment per kilowatt hour (kWh) of energy than does mining. Although it consumes less

[3] Metal and ores have accounted for between 70% and 80% of the country's exports since 2005 (World Bank, 2020d). A total of 80% of Zambia's copper comes from four mines. Barrick Gold, a Canadian corporation, owns Lumwana mine. First Quantam Minerals, a Canadian corporation, runs Kansanshi mine. Glencore International, a British corporation headquartered in Switzerland, and First Quantam Minerals own Mopani Copper Mines. Vedanta, an Indian corporation, majority-owns Konkola Copper Mines, while the remaining fifth is owned by ZCCM, majority-owned by the Zambian government and listed in Lusaka and on Euronext.

[4] National installed capacity was 2411 MW in 2015 (Energy Regulation Board of Zambia, 2016).

Table 1.1 Manufacturing's share of electricity consumption dropped from 2015 to 2017 in Zambia

National electricity consumption by economic sector, 2014-2018

Sectors	2014		2015		2016		2017		2018	
	GWh	% share	GWh	% share	GWh	% share	GWh	% share	GWh	% share
Mining	5871	47.3%	6246	54.5%	5918	54.5%	6202	50.9%	6682	54.8%
Domestic	3251	26.2%	3482	30.4%	3383	31.2%	4147	34.0%	4337	35.6%
Finance & property	487	3.9%	517	4.5%	499	4.6%	640	5.2%	714	5.9%
Manufacturing	479	3.9%	**531**	**4.6%**	**470**	**4.3%**	503	**4.1%**	593	**4.9%**
Agriculture	241	1.9%	260	2.3%	228	2.1%	262	2.1%	297	2.4%
Others	99	0.8%	99	0.9%	80	0.7%	87	0.7%	84	0.7%
Trade	107	0.9%	110	1.0%	97	0.9%	110	0.9%	114	0.9%
Energy & water	73	0.6%	89	0.8%	88	0.8%	81	0.7%	69	0.6%
Quarries	62	0.5%	68	0.6%	60	0.5%	118	1.0%	148	1.2%
Transport	31	0.3%	33	0.3%	28	0.3%	32	0.3%	33	0.3%
Construction	1702	13.7%	15	0.1%	7	0.1%	10	0.1%	11	0.1%
Total	**12,405**	**100%**	**11,450**	**100%**	**10,857**	**100%**	**12,192**	**100%**	**13,080**	**107%**

Author's creation. *Sources*: Energy Regulation Board of Zambia (2016, p. 0, 2017, p. 9, 2018, p. 36, 2019, p. 39), Ahmed et al. (2020)

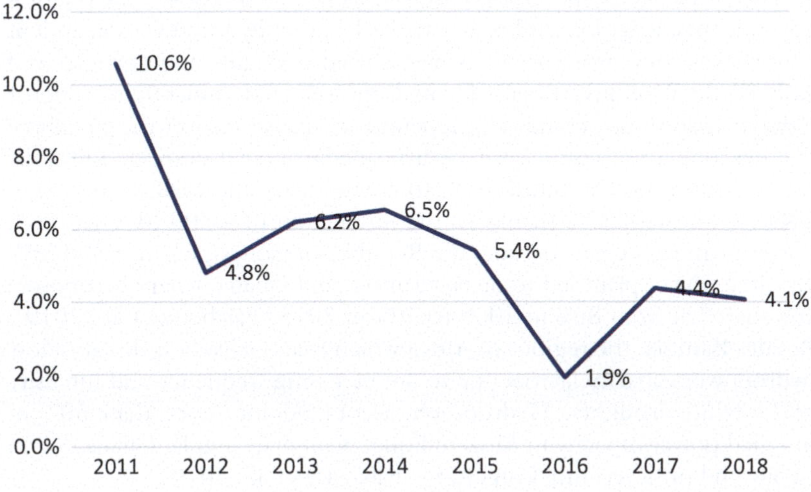

Fig. 1.1 Zambia's real manufacturing value-added growth declined following the power outages of 2015 and 2016. (*Data source*: World Bank, 2020c; graph: own)

than a tenth of energy that mining does, it contributes three-quarters of what mining does to GDP, and employs almost three times the workforce.[5] Second, focusing on manufacturing allows us to interrogate the political economy of funding for energy as well as Zambia's system of provision's path dependence and underlying economic assumption of allocative efficiency in charging subsidised tariffs that together resulted in, by 2015, underinvestment in diversified power generation. The investment rationale of supplying power for mineral extraction, beneficiation and industrialisation that holds for Zambia vis-à-vis the World Bank's history of prioritisation of hydropower can also be seen to directly apply to Brazil, Colombia, Ethiopia, Ghana, Kenya and Turkey. Further, because the World Bank's early appraisals of hydropower 'rarely' investigated the 'possibility of a thermal alternative' (Mason & Asher, 1973, p. 237), countries receiving hydropower investment were set on a trajectory of building power generation infrastructure that would later prove not to be climate resilient.

This work builds upon the commission's study and assesses the political economy of energy in Zambia which like 17 other hydropower-dependent countries is now prone to El Niño-induced droughts and also received early World Bank investments for hydropower predominantly to support industrialisation or mining (see Appendix A). In Africa over the past three decades, only some countries have diversified from hydropower: Tanzania, where hydropower's contribution to energy generation fell from 95% in 2000 to a low of 37% thanks to the installation of a 700 MW gas-fired power plant in the past decade; the Republic of Congo, where a 300 MW gas-fired power plant led to diversification; and Ghana, where hydropower's share fell from 80% in 2000 to 50% in 2015 (Falchetta et al., 2019). Besides Zambia, the regions of Africa which face the likeliest threat of low hydropower capacity factors due to an increasing frequency and intensity of El Niño droughts. Hydropower accounted for more than 50% of installed power capacity in Mozambique, Namibia, Angola, Ghana, Sierra Leone and until recently Cameroon (Falchetta et al., 2019).

[5] Mining, which consumed more than half of Zambia's electricity, contributed to 11.2% of GDP less net taxes, compared with manufacturing which consumed 4.3% of electricity but contributed to 8.5% of GDP less net taxes (CSO, 2018). Manufacturing employs 8.1% of employed people while mining employs 2.9% (CSO and Ministry for Labour and Social Security, 2019). The Central Statistical Office (CSO) statistics for manufacturing include food, beverages, tobacco and textiles which the World Bank's definition exclude (CSO, 2018; World Bank, 2020b).

This work lays out the historical, political and economic forces that saw the propagation of hydropower across the developing world; investigates why Zambia had no diversity in its energy production portfolio prior to 2015; evaluates the implications of an undiversified energy mix to Zambia's manufacturing firms, as well as to the environment; reviews options for a 'least regret' energy mix;[6] and concludes with recommendations. In so doing, it attempts to answer the following research questions:

1. How did Zambia come to be so reliant on hydropower?
2. What are the impacts of this reliance on Zambia's industrialisation, how do manufacturers respond, and how effective are their responses?
3. How does Zambia avoid power outages and its associated costs in the future in a cost-effective and minimally destructive manner with respect to the climate crisis?

The originality and contribution to knowledge of this work stands on

- an incremental expansion of the historical institutional path dependency model, fusing together the way in which Mahoney, Pierson and Skocpol, and Stinchcombe (Stinchcombe, 1968; Mahoney, 2000; Pierson & Skocpol, 2002) model it and building upon their combined framework;
- bringing new insight into how Zambia's system of provision of energy became locked into what Liebowitz and Margolis (Liebowitz & Margolis, 1995) classify as
  - first-degree path dependence (where sensitivity to starting points exist but do not result in inefficiency) with respect to choice of energy mix for investment decisions made prior to the 1950s;
  - second-degree path dependence (the inferiority of a chosen path that is unknowable at the time a choice was made) for investment decisions made in the 1970s; and

[6] It appears as though use of the term 'least regret' with respect to future energy scenarios started gaining traction in 2015/2016 (National Grid, 2015; Sanders et al., 2016; Zachary, 2016). I first heard its use at the Energy and Economic Growth Grid Reliability and Utility Operations conference in February 2020 in Accra, Ghana.

  - third-degree path dependence for investment decisions made in the 2010s, where dependence is based on initial conditions and ignores information (about the increase and frequency of El Niño-induced droughts, and the competition for water that exists between hydropower and food and water) which implies that it is inefficient.
- the primary data collected through
  - surveys with manufacturing firms to bring fresh insight into the impact of path dependence on electricity consumers;
  - interviews to contextualise and triangulate interpretations with key informants from
    - within the Zambian government and ZESCO,
    - among private energy developers,
    - among the interface between Zambian manufacturers and the government,
    - within a donor,
    - among Zambian researchers,
    - among other sources; and
  - unique analysis of both the primary and secondary data.

Thanks to the International Growth Centre research grant, this study was able to hire and project manage a team of eight in Zambia to collect 123 surveys (a third of the population of large Zambian manufacturing firms) from a representative sample of large manufacturing firms located in Zambia's main industrial hubs of Lusaka, Ndola and Kitwe. The objectives of the survey were to inform what the costs were of power outages to large manufacturing firms, and to inform what coping mechanisms firms employed (with the underlying hypothesis that backup diesel generators are at the forefront of such coping mechanisms). The survey sought to reveal the extent to which diesel generators are successful in mitigating firms' costs, and what costs they in turn bear on the environment. It further sought to identify predictors of firms' propensity to invest in backup generator capacity, firms' propensity to use that installed capacity, firms' carbon emissions from backup generation, and to estimate the extent of emissions from diesel generation by the manufacturing sector.

This study finds that because of the path dependent lock-in to hydropower, Zambia's system of energy provision did not adapt to warnings given as early as 2001 (IPCC, 2001) that hydropower in the Zambezi River Basin was no longer going to be the dependable source of baseload power generation that it once was due to climate change. The increase in global temperatures was forecast to increase the frequency and intensity of drought and low reservoir levels and therefore hydropower capacity factors (Harrison & Whittington, 2002; Beilfuss, 2012; Spalding-Fecher, 2012). Further, with the lock-in into hydropower, Zambia's system of grid electricity provision was going to be increasingly competing with food and water. The lock-in of low tariffs to consumers further exacerbated the issue by financially constraining ZESCO's ability to contract the additional required power generation. ZESCO had the opportunity to respond to the threat of El Niño droughts with greater urgency when it was forced to negotiate by the Government of Zambia with the independent power producer for the Maamba Collieries coal power plant, but ZESCO delayed contracting a non-hydropower source of energy first by stalling on the tariff negotiation. A new power plant's capital expenditures would not have been fully amortised, compared to the majority of ZESCO's power-generating assets, which had for decades allowed ZESCO to charge a tariff to consumers lower than what it would have cost ZESCO to purchase power from an independent power project (IPP). Second, ZESCO stalled by delaying the construction of infrastructure to connect the power plant with its grid. ZESCO would have treated diversification away from hydropower as urgent had ZESCO not been locked into seeing only the merit of hydropower generation and had ZESCO not been locked into a mindset of being able to continue charging low tariffs to consumers. Maamba Collieries would have started delivering energy to the grid in 2015, thus preventing the power outages of 2015 and 2016, which adversely impacted Zambia's manufacturing firms.

The work is structured as follows.

Chapter 2 provides the context for how Zambia's system of energy provision came to be as it was. It describes the underlying economic theory and empirical evidence connecting energy and economic development through the medium of industrialisation. It introduces the agents within Zambia's system of provision, and explains the importance of the World Bank, which had financed power generation assets that had generated 81% of Zambia's energy in 2014 right before the power outages of 2015 and 2016. It then goes into the political context of the World Bank's decision-making, as well as the economic rationale explained by its economists and top executives. Chapter 2 then rounds out with a look at how the provision of energy impacted growth of the manufacturing sector in Zambia, and how successfully it contributed overall to development. In spite of manufacturing's inability to materially affect structural employment away from subsistence agriculture prior to 1991, its resurgence since the 2000s has seen it return more employment than the mining sector and more GDP per kilowatt hour of energy consumed.

With the threat power outages pose to this resurgence, Chap. 3 is a review of the literature with initial desk analysis on the impact of power outages on manufacturing.

In Chap. 4 we will look at how the power outages of 2015 and 2016 impacted Zambia's manufacturing. From a representative sample of 123 large manufacturing firms, Chap. 4 provides fresh insights into the impact of power outages on Zambia's manufacturing firms. It ranks damages incurred from outages by manufacturing firms, ranks their use of mitigation strategies, shows how the costs of power outages differ for firms using differing coping strategies and identifies firm characteristics that can be used to predict differing coping mechanisms. We will also identify firm behaviours that are suboptimal for grid efficiency, but also identify ways in which firms are willing to give back to the grid.

In Chap. 5 we will seek to understand why Zambia's system of energy provision did not prevent the power outages of 2015 and 2016. First, we will note the rising incidence of transmission and distribution losses. We will then look at Zambia's forecast demand for grid electricity to assess the additional power generation Zambia will require net of reducing transmission and distribution losses. We then review the literature Zambia's system of energy provision had at its disposal through the 2000s prior to the power outages of 2015 and 2016 on how vulnerable additional investment into hydropower in the Zambezi River Basin would be to climate change. Given that the system of provision proceeded to invest in hydropower in the Zambezi River Basin anyway, we will then see how the historical institutional path dependence framework explains this. We will also see how a path-dependent industrial strategy, together with electoral politics, contributes to our understanding as to why the system of provision has not sought a more equitable tariff structure and distribution of grid energy.

Chapter 6 offers supply- and demand-side recommendations for the agents in Zambia's system of provision. Chapter 7 zooms back out to the broader applicability of the case study of Zambia to low-carbon policies that ensure energy security for a fifth of the world's population.

References

Ahmed, I., Parikh, P., Sianjase, G., & Coffman, D. (2020, December 4). The impact decades long dependence on hydropower in El Niño impact-prone Zambia is having on carbon emissions through diesel backup generation. *Environmental Research Letters, 15*(12) https://doi.org/10.1088/1748-9326/abb6a1

Bayliss, K., & Pollen, G. (2019). *From Global to Local: Exploring the effects of the global energy policy paradigm in Zambia.* 119. Leeds. https://sri-working-papers.leeds.ac.uk/wp-content/uploads/sites/67/2019/10/SRIPs-119.pdf

Beilfuss, R. (2012). *A risky climate for Southern African Hydro: Assessing hydrological risks and consequences for Zambezi River Basin Dams.* International Rivers. http://seine.oieau.fr/oieau.fr/sites/seine.oieau.fr.oieau.fr.eaudoc/files/32987.pdf

Conway, D., et al. (2015, September). Climate and southern Africa's water–energy–food nexus. *Nature Climate Change.* Nature Publishing Group, 5, 837–846. https://doi.org/10.1038/nclimate2735

CSO. (2018). *Data portal tables.* Central Statistical Office of Zambia.

CSO and Ministry for Labour and Social Security. (2019). 2018 Labour Force Survey Report. *Ministry of Labour and Social Security*, p. 153. http://www.statistics.gov.my/portal/download_Labour/files/labour_force/Labour_Force_Survey_Report_Malaysia_2012.pdf

Energy Regulation Board of Zambia. (2016). *Energy Sector Report 2015.* http://www.erb.org.zm/reports/EnergySectorReport2014.pdf

Energy Regulation Board of Zambia. (2017). *Energy Sector Report 2016.* https://doi.org/10.13140/RG.2.1.3893.7043

Energy Regulation Board of Zambia. (2018). *Energy Sector Report 2017.* http://www.erb.org.zm/downloads/reports/esr2017.pdf

Energy Regulation Board of Zambia. (2019). *Energy Sector Report 2018.* Lusaka. Available at: http://www.erb.org.zm/reports/esr2018.pdf

Falchetta, G., et al. (2019). Hydropower dependency and climate change in sub-Saharan Africa : A nexus framework and evidence-based review. *Journal of Cleaner Production, 231*, 1399–1417. https://doi.org/10.1016/j.jclepro.2019.05.263

Harrison, G. P., & Whittington, H. W. (2002). Susceptibility of the Batoka Gorge hydroelectric scheme to climate change. *Journal of Hydrology, 264*(1–4), 230–241. https://doi.org/10.1016/S0022-1694(02)00096-3

International Growth Centre. (2018). *About—IGC, www.theigc.org*. Retrieved November 22, 2018, from https://www.theigc.org/about/
IPCC. (2001). *Climate change 2001: Impacts, adaptation, and vulnerability.* https://www.ipcc.ch/site/assets/uploads/2018/03/WGII_TAR_full_report-2.pdf
Liebowitz, S. J., & Margolis, S. E. (1995). Path dependence, lock-in, and history. *Journal of Law, Economics, and Organization, 11*(1), 205–226. https://doi.org/10.2139/ssrn.1706450
Mahoney, J. (2000). Path dependence in historical sociology. *Theory and Society, 29*(4), 507–548. http://www.jstor.org/stable/3108585
Mason, E. S., & Asher, R. E. (1973). *The World Bank since Bretton Woods.* The Brookings Institution.
Mwila, A., Sinyenga, G., Buumba, S., Muyangwa, R., Mukelabai, N., Sikwanda, C., Chimbaka, B., Banda, G., Nkowani, C. & Bwalya, B. (2017). Impact of load shedding on small scale enterprises. [online] Lusaka: *Energy Regulation Board.* Available at: [Accessed 22 June 2021].
National Grid. (2015). *Electricity capacity report.* https://doi.org/10.1016/j.matchemphys.2007.04.026
Pierson, P., & Skocpol, T. (2002). Historical institutionalism in contemporary political science. In Katznelson, I. and Milner, H. (eds.), *Political Science—State of the Discipline*, pp. 693–721. American Political Science Association and WW Norton & Company.
Sanders, D. et al. (2016). *An analysis of electricity system flexibility for Great Britain.* https://www.gov.uk/government/uploads/system/uploads/attachment_data/file/568982/An_analysis_of_electricity_flexibility_for_Great_Britain.pdf
Spalding-Fecher, R. (2012). *Powering Southern Africa—The future of hydro in the Zambezi River Basin, Climate & Development Knowledge Network.* Retrieved June 6, 2020, from https://cdkn.org/2012/02/powering-southern-africa-the-future-of-hydro-in-the-zambezi-river-basin/?loclang=en_gb
Stiglitz, J. E. (2017). Promoting Sustainable Industrial Policies. In *Industrialize Africa: Strategies, Policies, Institutions and Financing*, pp. 11–15 African Development Bank Group. https://www.afdb.org/en/news-and-events/industrialize-africa-strategies-policies-institutions-and-financing-17570/
Stinchcombe, A. L. (1968). *Constructing social theories.* University of Chicago Press.
Thurlow, J., Zhu, T., & Diao, X. (2012). Current climate variability and future climate change: Estimated growth and poverty impacts for Zambia. *Review of Development Economics, 16*(3), 394–411. https://doi.org/10.1111/j.1467-9361.2012.00670.x

Wang, G., et al. (2017). Continued increase of extreme El Niño frequency long after 1.5°C warming stabilization. *Nature Climate Change, 7*(8), 568–572. https://doi.org/10.1038/NCLIMATE3351

World Bank. (2019). *World development indicators: Zambia*. World Bank. https://data.worldbank.org/country/zambia

World Bank. (2020a). *Electricity production from hydroelectric sources (% of total)*. https://data.worldbank.org/indicator/EG.ELC.HYRO.ZS?view=chart

World Bank. (2020b). *Metadata glossary, databank*. Retrieved October 24, 2020, from https://databank.worldbank.org/metadataglossary/wdi-database-archives-(beta)/series/NV.IND.TOTL.CD#:~:text=Longdefinition,electricity%2Cwater%2Candgas

World Bank. (2020c). *Projects in Bolivia, World Bank*. Retrieved October 17, 2020, from https://projects.worldbank.org/en/projects-operations/projects-list?lang=en&searchTerm=&countrycode_exact=BO

World Bank. (2020d). *Zambia world development indicators*.

Zachary, S. (2016). *Least worst regret analysis for decision making under uncertainty, with applications to future energy scenarios*. Edinburgh. http://arxiv.org/abs/1608.00891

ZICTA and CSO. (2018). *2018 National survey on access and usage of information and communication technologies by households and individuals—A demand side assessment of access and usage of ICTs in Zambia*. https://www.zicta.zm/Downloads/publications/2018ICTSurvey_PreliminaryReport.pdf

CHAPTER 2

Energy, Industrialisation and Economic Growth

In this chapter we explore the theory justifying supplying energy supply for manufacturing by linking manufacturing to economic development; the agents in Zambia's system of energy provision and the importance of the World Bank among them; the underlying political interests and economic rationale within the World Bank for promoting hydropower and exporting the Tennessee Valley Authority template to low-income countries; and how application of the TVA template played out in Zambia. We will seek to understand how energy supply and other factors gave rise to Zambia's industrialisation, de-industrialisation and reindustrialisation to then set the scene in Chap. 3 for the more recent threat of power outages on Zambia's industrialisation.

2.1 Mechanisms by Which Industrialisation Drives Economic Growth

Economic prosperity and electricity consumption have gone hand-in-hand. The log of electricity consumption is positively correlated with the log of countries' gross domestic products (Castellano et al., 2015, Fig. 1). One, however, does not directly cause the other.[1] Rather, causality goes both ways through intermediate steps. On the one hand, with greater

[1] Dlamini et al. found no causal relationship between the two variables using full-sample Granger causality tests (Dlamini et al., 2016).

I. Ahmed, *The Political Economy of Hydropower Dependant Nations*, https://doi.org/10.1007/978-3-030-71266-2_2

prosperity, consumers move up the energy ladder and purchase more goods such electric cookers, refrigerators and water heats (Tobey, 1996, pp. 122, 157–162). The consumption in turn drives sales of appliances which need to be manufactured, which takes energy. Increased manufacturing means increased jobs, which means increased disposable income to buy further goods which need to be purchased. This is the Keynesian multiplier effect that was observed with the availability of cheap electricity with the creation of the Tennessee Valley Authority as we show below. Likewise, electricity availability enables increases in economic productivity—from that at the household level (Ahmed, 2020a, p. 196) to enabling large-scale industrialisation which involves moving labour into higher productivity sectors as the cited literature below elucidates. The increased productivity drives economic growth. Kaldor's first law of growth posits that the higher the growth of manufacturing output, the more significant is the growth rate of the economy's product as a whole (Marconi et al., 2016). Kaldor's second law, also known as the Kaldor-Verdoorn law (Marconi et al., 2016), states that with increasing returns to scale and increasing skill and know-how, opportunities for easy communications of ideas, increasing opportunities for differentiation of processes and specialisation of activities increase such that productivity grows proportionally to the square root of output (Kaldor, 1970, p. 484).

Szirmai and Verspagen (2015) found empirically that Kaldor's hypothesis that manufacturing is an engine of economic growth held: manufacturing had a moderately positive impact on economic growth for a panel of 88 countries for the period 1950–2005. Cantore et al.'s (2017) regression analysis controlling for endogeneity bias using a sample of 80 countries for 1980–2010 similarly supported Kaldor's hypothesis.

Several theoretical arguments contribute to understanding why manufacturing has empirically been found to have historically propelled economic growth:

1. Manufacturing is more capital intensive than agriculture, and thus provides more opportunities for investment (Haraguchi et al., 2019; Szirmai, 2012).
2. Being capital intensive, manufacturing encourages investment into technological advance to reduce costs through achieving economies of scale. Technological advance is also driven to offer customers unique products, and thus provide the firm economic rents. Technological advance allows an economy's production capabilities

to increase and hence its potential income, according to the neoclassical Solow model (Solow, 1956).

3. These technological advances have a higher likelihood of spilling over into other sectors of the economy thanks to manufacturing's strong backward and forward linkages with other sectors. Manufacturing uses primary sector goods, and generates demand for tertiary sector services (Kaldor, 1984). Szirmai (2012) argues that advances in ICT hardware technologies produced in the manufacturing sector fuelled technological advance in the software-producing service sector. (Thurlow & Wobst, 2007, pp. 232, 234) show that Zambia's services sectors grow more rapidly in the industry-led scenario. They note, however, that agriculture has more pro-poor growth linkages.
4. Given manufacturing's superior use of technology and capital, a structural transfer of resources from the primary sectors to manufacturing results in a structural dividend due to increased productivity. A transfer of unskilled labour from low-productivity agriculture to high-productivity industry results in an immediate increase in overall productivity per capita and has been a major source of growth in developing countries (Rodrik, 2016; Szirmai, 2012). Rodrik (2016) finds that formal manufacturing exhibits unconditional labour productivity convergence. Cantore et al.'s (2017) research points to manufacturing productivity and manufacturing employment share as being responsible for observable effects on growth.
5. Given the higher productivity per worker, a burgeoning manufacturing sector not only absorbs labour, but also should be better able to compensate workers.
6. As per capita income rises both as resulting from and as resulting in increased manufacturing, demand for manufactured goods increases and the share of agricultural expenditures declines (Engel's law). Economies void of manufacturing will not profit from rises in per capita income, while economies with manufacturing sectors that can survive disruptive changes and international competition can benefit from the Keynesian multiplier.
7. Manufacturing goods are easily tradable, as opposed to services, thus allowing a sector within a developing economy to escape the constraint of low-income domestic demand (Rodrik, 2016), and again allowing for the Keynesian multiplier to amplify the benefits of a thriving manufacturing sector.

8. When trade opens between a country that is more industrialised and another that is less industrialised, the more industrialised nation will be able to supply the needs of the more agricultural country on more favourable terms, and indeed will eliminate its competition for demand in the foreign market (Kaldor, 1970, p. 484).

In summary, electrification enables industrialisation which has historically led to economic development in industrialised economies. Since electricity consumption is not simply the result of individual preferences but connected to the way in which it is provided (Bayliss & Pollen, 2019), we will now interrogate and build upon Bayliss and Pollen's model of Zambia's energy system of provision which enunciates the agents and their roles in Zambia's supply of electricity.

2.2 Zambian Energy's System of Provision

Bayliss and Pollen's 'system of provision' approach to analysing Zambia's energy provision contends that consumption outcomes are not simply the result of individual preferences but are inherently linked to production, rooted in local context and emerge from the tensions among agents in the chain of provisioning (Bayliss & Pollen, 2019). The approach recognises the roles and significance of a diverse body of agents.

Figure 2.1 incorporates Bayliss and Pollen's system of provision (Bayliss & Pollen, 2019, fig. 1). The power utility ZESCO, independent power projects (IPPs) that sell energy to ZESCO through power purchase agreements and the Southern African Power Pool (SAPP) generate power. This power is purchased by ZESCO, an autonomous corporation established in 1969 and wholly owned by the Government of the Republic of Zambia (IBRD and IDA, 1973, p. i) which then distributes energy to end-consumers (households, businesses and some mines) as well as to the Copperbelt Energy Corporation (CEC), an intermediate power utility which distributes solely to mining companies in the Copperbelt Region and also sources energy from the Southern African Power Pool. Overseeing energy policy is the Ministry of Energy. The Rural Electrification Authority is its agency to oversee and implement rural electrification policy. The Energy Regulation Board is responsible for licensing, tariff-setting and regulating quality of supply using a series of key performance indicators. Bayliss and Pollen (2019) assert that the tariffs paid by the Copperbelt Energy Corporation and mines are outside the remit of the Energy Regulation Board. In fact, the remit of the Energy Regulation Board is

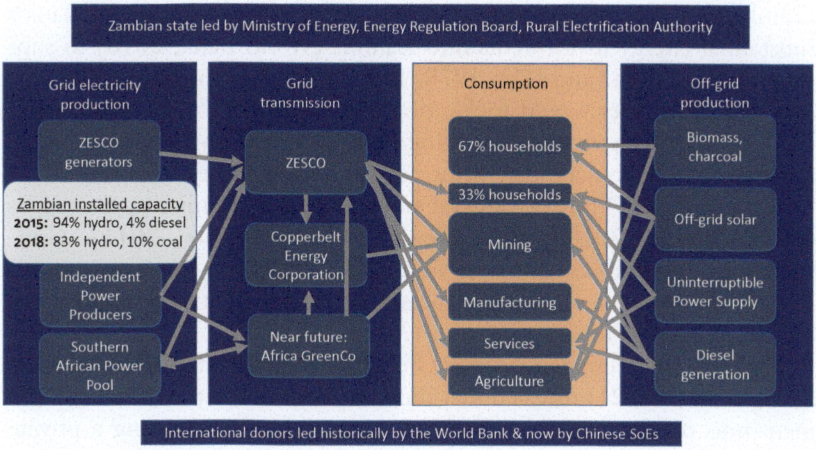

Fig. 2.1 Agents in the system of provision for electricity or electricity substitutes in Zambia. (Source: Ahmed et al. (2020))

currently being litigated (personal interview with Alfred Mwila, Director of Economic Regulation at the Energy Regulation Board, 6 June 2018, and public statement made by Cletus Sikwanda, Economist at the Energy Regulation Board on 4 September 2019 at a public forum in Lusaka organised by the International Growth Centre to present some of this research's findings; Ahmed, 2021). Parliament has the power to resolve this ambiguity (personal interview with Alfred Mwila, Director of Economic Regulation at the Energy Regulation Board, 6 June 2018; Ahmed, 2021), and so Parliament can also be considered an agent in the System of Provision.

Figure 2.1 builds upon Bayliss and Pollen's system of provision by including off-grid electricity provision, by taking into account the possible entry of a prospective private off-taker and by recognising the World Bank's pre-eminent role among donors and investors in financing Zambia's power generation assets.

The modified system in Fig. 2.1 also includes the 67% of the population that live without grid connections (ZICTA and CSO, 2018) and who rely on off-grid solutions including solar products, charcoal and biomass

(Tembo, 2018).[2] This situation persists 50 years after the World Bank's substantial energy investments into Zambia (World Bank, 2018) to support mining (IBRD and IDA, 1970, 1973) in the vain hope that the benefits of hydropower would trickle down (Park, 1968). Agricultural firms can meet their electricity needs using biomass generation (Zambia Sugar, 2017, p. 15). Manufacturing firms use backup diesel generation to make up for lost ZESCO power (Ahmed et al., 2019), as do mining companies (Mfula, 2010a; 2010b; Steel News, 2011; Syndigate, 2011). Grid-connected households' favourite mitigation strategy to power outages is rechargeable lights and solar products; the 10th and 11th favoured strategies are UPSs and diesel generators (Ahmed, 2020b).

Personal interviews with Africa GreenCo (23 April 2019, 3 September 2019; Ahmed, 2021) revealed that the company intends to facilitate investment into renewable independent power projects by becoming a private credit-worthy off-taker that on-sells to ZESCO and clients on the Southern African Power Pool. Africa GreenCo's *raison d'être* is to facilitate new renewable energy investment that would not otherwise take place because prospective independent power projects are afraid that ZESCO will not pay them because ZESCO is not credit-worthy (phone interview with a donor agency energy expert, 27 February 2019). Africa GreenCo has raised capital from donors and development finance institutions to provide it 18 months working capital buffer in order to be able to provide this service (personal interviews with the CEO and CCO on 23 April 2019 and 1 August 2019; email received from CEO 27 October 2020; Ahmed, 2021).

Without the World Bank and its regional counterpart, the African Development Bank, very little large-scale infrastructure would have existed in Africa because of the lack of established markets and lack of trust by private sector participants in both their governance and commercial demand. The World Bank, AfDB and other development finance institutions both stand in place of as well as catalyse private investment by issuing guarantees. More recently, the Government of China has been investing heavily into African infrastructure as it joins the West's race for African resources (Moyo, 2013). Indeed, Chinese state owned enterprises are ZESCO's largest current debtor, with debts to Chinese institutions ranking first, second, seventh, 12th, 17th, 21st and 23rd, compared to the

[2] In the context of rural China, Jiang et al. (Jiang et al., 2020) found that in low-income households, the main sources of energy increase, the consumption of energy increases and the energy flows become more frequent.

World Bank's loans ranking 9th, 14th, 24th and 29th largest debts as of 2018 (Zesco Ltd, 2018, p. 98).

Even though its most significant investments into Zambian power generation had taken place 40 years prior, the World Bank was still Zambia's most significant investor in Zambian power generation assets in use prior to the power outages of 2015 and 2016. Since these hydropower dams were completed, hydropower accounted for almost all of Zambia's power supply (see Fig. 2.2). In 1971, as far back as World Bank panel data go back, hydropower accounted for 75% of Zambia's power generation. This increased to 92% in 1972 and to 98% by 1977. From 1982 to 2013 inclusive, this proportion did not fall below 99% (World Bank, 2020f). In 2014, the Energy Regulation Board observed more than 95% generation from hydropower (Energy Regulation Board of Zambia, 2015, p. 3), with the balance provided by diesel, heavy fuel oil and a minimal amount of solar photovoltaic plants.

Just before the power outages of 2015 and 2016 in 2014, the Kariba and Kafue hydropower projects that the World Bank financed between 1956 and 1974 accounted for 81% of ZESCO generated electricity (Energy Regulation Board of Zambia, 2015, pp. 4, 5, 6, 7). Since that time and up until the power outages of 2015 and 2016, no significant investment had been made in Zambia's power generation capacity (Whitworth, 2014, p. 14). The World Bank not only financed the Kariba and Kafue hydropower projects, but, in the words of the former chairman

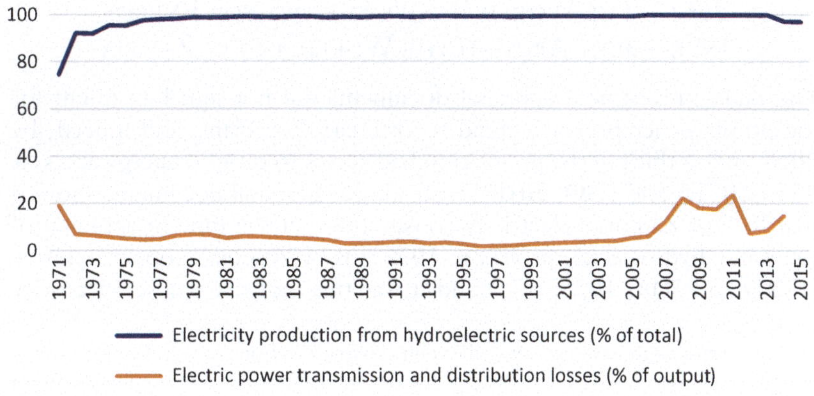

Fig. 2.2 Zambia's historic reliance on hydropower. (Data source: World Bank, 2018; graph: own)

Table 2.1 Zambia's first 18 executed World Bank-financed infrastructure projects

#	Project Title	Project ID	Commitment, USD M current	Status	Approval Date	US CPI-U	Power commitment, USD M constant 9/2018	Economic infrastructure, USD M constant 9/2018	Bottom of the pyramid focus, USD M constant 9/2018	
18	POWER KARIBA-NORTH(S	P003166	42.10	Closed	05/07/1974	49.4	215	215		
17	Lusaka Squatter Upgrading and Sites and Services Project	P003167	20.00	Closed	05/07/1974	49.4			102	
16	Kafue Hydroelectric Power/Second Stage Project	P003164	115.00	Closed	05/07/1973	44.3	655	655		
15	Program Loan Project	P003163	30.00	Closed	14/06/1973	44.2			171	
14	Education Project (03)	P003162	33.00	Closed	31/05/1973	43.9			190	
13	Integrated Family Farming Project	P003161	11.50	Closed	16/01/1973	42.6			68	
12	Kariba North Power Project	P003160	40.00	Closed	07/07/1970	39.0	259	259		
11	Agriculture / Tobacco Project	P003158	5.50	Closed	28/05/1970	38.6		36		
10	Education Project (02)	P003159	5.30	Closed	18/11/1969	37.0			36	
9	Livestock Project	P003154	2.50	Closed	17/06/1969	36.6		17		
8	Education Project (01)	P003156	17.40	Closed	08/04/1969	36.3			121	
7	Forestry Project	P003155	5.30	Closed	24/09/1968	35.1			38	
6	Highway Project (02)	P003157	10.70	Closed	24/09/1968	35.1		77		
5	Highway Project (01)	P003153	17.50	Closed	04/10/1966	32.7		135		
4	POWER II / Kariba Electric Power Development Project (02)	P003152 / P003267	3.80	Closed	02/10/1964	31.1	31	31		
3	Railway Improvement Project	P003151	9.50	Closed	16/06/1958	28.9		83		
2	POWER I / Kariba Electric Power Development Project (01)	P003150 / P003264	40.00	Closed	21/06/1956	27.2	371	371		
1	Railway Project (01)	P003149	14.00	Closed	11/03/1953	26.9		131		
					Total investments, 1953-74:		1,531	2,011	727	2,738
					% of total investments,1953-74:		56%	73%	27%	

Sources: World Bank, 2018 https://projects.worldbank.org/ and US Bureau of Labor Statistics, 2018 Table 24 for CPI-U values

of the British construction company selected to build the Kariba Dam, 'twisted' the Zambian government's arm to make it accept the Kariba North Bank power station (Morrell, 1987, pp. 152–153).[3] Similarly, power generation was the World Bank's most significant destination for investment in Zambia, accounting for more than half the value of its investments up until 1974 (see Table 2.1). It therefore behoves us to understand the dynamics and the thinking that motivated and informed the Bank's investment prioritisation of energy.

2.3 The World Bank's Support for Energy Investment Around the World and in Zambia

The Bank's investment appraisal documents did not signal an intent for the power generation to extend to ordinary Zambians, and indeed, by 2018, only a third of the population had access to on-grid energy access is 33% (ZICTA and CSO, 2018). Instead, the appraisal documents showed that 85% of Zambia's electricity consumption at the time of appraisal of the Kafue hydroelectric project in 1973 was by Zambia's copper industry (IBRD and IDA, 1973, p. 1) and it was this industry whose 'electricity

[3] The twisting of the arm was not in the context of the Zambian power utility not wanting more hydropower—it did. The twisting of the arm was in the context of helping the World Bank finance the hydropower project for the benefit of the former British colony Southern Rhodesia, which was facing sanctions, as well as Zambia, by channelling the money through Zambia (Morrell, 1987).

requirements' (IBRD and IDA, 1973, p. 1) were being assessed. Consistent with this neglect of electricity access for all as a priority, only a quarter of the World Bank's investments by real value were directly targeted at 'social infrastructure' to address poverty reduction in Zambia and the requirements of Zambians (see Table 2.1).[4]

The Bank's prioritisation of economic infrastructure, and moreover hydropower, reflected both the political instrumentalisation of the Bank within a geopolitical context and its institutionalised economic thinking. To understand the Bank's political instrumentalisation as well as the investment principles underlying particular investments, we need to look at the original template of the hydropower project (the Tennessee Valley Authority), to the context which birthed the Bank, rhetoric that guided its purpose, as well as the writings and sayings of is most influential economists and executive leadership in respect of power generation versus investment in more welfare-focused projects.

2.3.1 *The Tennessee Valley Authority Template*

The Tennessee Valley Authority was a federal hydropower agency that would by 1946 comprise nine 'huge' dams along a stretch of 600 miles of Tennessee River, and a further 11 dams on the tributaries. As well regulating the flow of water, the dams had the capacity to generate 2–2.5 gigawatts of power, ranking the TVA second among US electric systems (Huxley, 1946, pp. 12–13). As it started delivering electricity in 1937, the Tennessee Valley Authority offered electricity prices that were almost 60% lower than the national average (Finer, 1972, p. 205). Whether or not prices converged as a result of competitive pressure, that fell to 45% by 1941 (Finer, 1972, p. 205; see Table 2.2 below). Where previously electric utilities had targeted one-fifth of households already modernised, the TVA focused on the four-fifths of households that utilities claimed were incapable of modernisation.[5]

[4] While the World Bank today no longer persists with the distinction between economic and social infrastructure (though the OECD does persist (OECD, 2019)), it is important to note the World Bank did make this distinction at the time that it financed most of Zambia's existing energy infrastructure. As late as 1995, the World Bank categorised water supply and sanitation as 'social' investments, separate from 'infrastructure' which were telecommunications, transportation, electric power and other energy (Kapur et al., 1997, p. 6).

[5] This was the key difference between the balanced growth targeted by the TVA and the unbalanced growth which ZESCO targeted: the TVA was for those excluded; ZESCO was for those already included. We will return to this later.

Table 2.2 The Tennessee Valley Authority's tariffs compared to the national average

	Average rate per kWh/cents					
Electric system	*1933*	*1937*	*1938*	*1939*	*1940*	*1941*
Tennessee Valley Authority		1.83	1.95	2.16	2.06	2.05
US entire industry	5.49	4.39	4.14	4	3.84	3.73

Source: Finer, [1944] 1972, p. 205 (author's adaptation); Ahmed (2020a)

Prior to the New Deal, the average household consumed 30 kWh monthly in the United States, just more than lights (11 kWh/month) and a radio (7 kWh/month). By the 1950s, that had increased by five times. Refrigerators—which spent on average 22 kWh/month—went from being a luxury-appliance to a mass appliance (Tobey, 1996, pp. 157–162). Greater demand from the middle-class market meant that their unit cost of production fell in addition to the cost of running them thanks to the TVA. Within three years of connecting to the TVA's Wilson Dam, Tupelo, a town of 10,000 people in Mississippi, saw eight stores sell USD 500,000 worth of appliances. More broadly, between 1933 and 1939, nominal retail and service receipts increased by 87% in the power area of the Tennessee Valley Authority according to census of business data (Finer, 1972, p. 213). A 1935 edition of *Electrical World* noted that the rise in sales of appliances was stimulated by the TVA's lower electricity rates (Tobey, 1996, p. 122). Electric cooking (123 kWh/month) and water heating (168 kWh/month) were the next items on the US energy ladder (Tobey, 1996, pp. 157–162).

Greater aggregate demand led to greater output and output per capita, and this was particularly pronounced in the TVA's catchment area. Per capita increased by 84% from 1933 to 1940 in Alabama, Mississippi and Tennessee, compared with 57% for the United States (Finer, 1972, p. 212).

The increase in sales explains the increase of 17,000 jobs in the trade sector in the TVA power area from 1930 to 1940 (Finer, 1972, p. 211). The greater prosperity in this sector explains the expansion of other sectors. Value-added by manufacturing increased by 39% from 1935 to 1939 in the TVA public power area (compared with 31% across the United States), and manufacturing wages increased by 35% (compared with 25% across the United States) (Finer, 1972, p. 211). The textiles, food products and chemicals manufacturing sectors increased employment by 17,000 from 1930 to 1940 (Finer, 1972, p. 211). In turn, the expansion of these sectors explains the increase in employment of the construction industry by 16,000 for the same period. All this despite the inference that the TVA power area imported its manufactured electrical appliances from outside the area.

Additionally, electrification's intended aim to curtail urbanisation was somewhat successful. The rural non-farm population of the TVA power area increased by 22% from 1930 to 1940, compared with a 17% increase for the urban population (Finer, 1972, p. 211).

More broadly, mass electrification to stimulate the economy and reduce unemployment redefined the role of government by defying and rewriting textbook teaching (Barber, 1996, p. 1). It provided the template for the Keynesian multiplier linking consumption and output in a cyclical relationship, and gave rise to the economic theories of a number of other Western economists, including Keynes, Rosenstein-Rodan and Hirschman, Rostow and Kuznets.[6] The latter two approached the question of economic development as a teleological exercise with respect to US political interests.

2.3.2 *The Bank's Political Instrumentalisation*

US President Frank Delano Roosevelt's administration designed the World Bank ('the Bank'), or the International Bank for Reconstruction and Development (IBRD) as it was conceived, as part of a suite of Bretton Woods organisations to help maintain peace after the Second World War (Mason & Asher, 1973, p. 15; Sharma, 2010, p. 33, 2017, p. 7) by providing international public goods in the form of monetary stability, free flow of capital and free trade (Gilpin, 2001). In 1945, President Roosevelt shared with Congress a vision of establishing institutions that complemented his other legacy of electrification through the Tennessee Valley

[6] Perhaps the most neglected underlying theoretical commonality between the mainstream British economist Keynes, Polish-naturalised British economist Rosenstein-Rodan and German-naturalised American economist Hirschman of the 1930s, '40s and '50s (the latter two of whom have latterly become known as development economists) was that their economic modelling based on the TVA experience was what Reinert (Reinert, 2019) characterises as *Erfahrungswissenschaft*—a science based on experience. Reinert is therefore not even-handed when he distinguishes German economics an *Erfahrungswissenschaft*' (Reinert, 2019).

Authority: 'expanded production, employment, exchange and consumption' to 'ensure our own prosperity [...] as well as for a peace that will endure' (Roosevelt, 1945). Prosperity the world over meant deeper markets with which the United States could trade, and fewer security threats. Roosevelt's vision would endure during the period in which the World Bank invested in Zambia's system of energy provision. Two decades later, US Secretary of Defence Robert McNamara who two years later would preside over the World Bank for the following 13 years, reiterated Roosevelt's vision by talking about the nexus between violence and 'economic backwardness', and that propelling developing nations from poverty was key to US security (McNamara, 1966).

With Roosevelt's successor's Truman Doctrine, the Bank and the Tennessee Valley Authority became instrumentalised to keep at bay the Soviet communist threat to US exports (Ahmed, 2020a). Providing cover to this mercantilism were the economic theories of British and American *Erfahrungswissenschaftler*[7] of the day whose work had been informed by the experience of the Tennessee Valley Authority—Keynes, Rosenstein-Rodan, Hirschman, Rostow, as well as Kuznets. The latter two approached the question of economic development as a teleological exercise with respect to US political interests.

The Bank's push for electrification to fulfil US interests rather than the best interests of the recipient countries fits Gilpin's global hegemon model of the United States, where the hegemon 'created a liberal international economy primarily to promote its own interests' (Gilpin, 2001, p. 99). Voting power in the Bank would be decided by shares held—rather than one country one vote as in the United Nations—making the United States, which holds more than twice the shares than the next largest shareholder (IBRD, 2020), the most influential country at the World Bank. The US Secretary of the Treasury sits on the World Bank's Board of Governors, the World Bank's highest governing body (World Bank, 2017b). All World Bank presidents have been American. It was left to the newly established regional development banks—the Inter-American Development Bank established in 1959, the African Development Bank in 1964, and the Asian Development Bank in 1966—to focus on poverty lending and water investments (Bakker, 2013, p. 288).

[7] Experiential scientists.

2.3.2.1 The Political Economy of Rostovian Growth and the Tennessee Valley Authority Template

Prime in intertwining development logic with the Truman Doctrine of containing Soviet expansion and influence was Walt Whitman Rostow's 'The Stages of Economic Growth: A Non-Communist Manifesto', published in 1960. According to the popular and influential modernisation theory of the American economist Walt Whitman Rostow who advised the Kennedy and Johnson administrations, economic development was a linear path of five stages (Rostow, 1960). (For an entertaining history of stages of economic growth dating back to antiquity, see Reinert (2019, pp. 289–319).

Acknowledging that the 'Keynesian revolution' (Rostow, 1960, p. 155) had allowed Western governments to address unemployment in times of economic depression, and thus thwart Marxism, he set about merging a dynamic version of classical production theory with Keynesian income analysis. Since underdeveloped nations were 'the main focus of Communist hopes', their take-off would be the 'most important single item on the Western agenda' (1960, p. 134).

Rostow's theory of production focused on the composition of investment and on developments within particular sectors of the economy at particular stages (1960, pp. 13–14)— at a 'traditional' stage; at a stage where it was getting ready for take-off and where 75% of the working force was in subsistence agriculture; at the stage of taking-off, by which end 40% of the labour force might work in agriculture; at a mature stage, where 20% of the work force would be in agriculture; and finally at a stage of high mass consumption (1960, p. 71).

Given the varying elasticities of demand with each stage of growth, Rostow prescribed a 'sequence of optimum patterns of investment' (1960, p. 14). Investment to increase societal productivity would thus start in agriculture; then in 'social overhead capital'.[8] Together, agricultural and infrastructure investment provided a 'viable base' (1960, p. 25) for modern industry. To maintain industrialised economies' growth, they should look towards 'underdeveloped' markets (1960, p. 156). Herein lies potential for a contradiction in the Rostovian development model: if industrialised nations are to continue growing by exporting to underdeveloped

[8] This was following Hirschman's use of the term in 1958 (p. 63) which Hirschman said included 'all public services from law and order through education and public health to transportation, communications, power and water supply, as well as such agricultural overhead capital as irrigation and drainage systems. The hard core of the concept can probably be restricted to transportation and power.' For his part, Rostow emphasised transportation.

markets, how will underdeveloped markets advance to the stage of mass high-consumption? Do they too not need to industrialise? If imports hinder domestic industrial development, which country's development should take precedence?

This potential conflict of interests was borne out in the World Bank's first development investment, with US commercial interests profiting from the investment.[9] In step with the clamouring in the United States for the 'weaponisation' of the Tennessee Valley Authority, the Bank's first development (as opposed to reconstruction) investment loan was made in hydroelectric projects in Chile from an array of options that included forestry, a harbour and transport (IBRD, 1947, 1948; Olivier, 1961, p. 5; World Bank, 2016). Power supply was deemed insufficient in Chile, and hydro was favoured over coal and liquid fuels because they were high in cost and insufficient in supply (IBRD, 1948, p. 22). This investment was followed, in line with Rostovian alertness for the need to establish export markets in underdeveloped markets, with a loan to Chile for the importing of agricultural machinery (World Bank, 2016, p. 4). In 1948, 32% of Chile's workforce was employed in agriculture and fishing (Díaz et al., 2016, p. 666), rendering it at Rostow's 'take-off stage' of development, or alternatively, ripe for flooding with US manufactured exports, required to sustain the final stage of growth in developed nations' high mass consumption. The ex-post data suggest that flooding with US manufactured exports is exactly what happened. UN (1962, p. 20) and World Bank (2016, p. 4) data show an 86% correlation between the value of Chile's manufactured imports and its increase in installed energy capacity between 1938 and 1960. Further analysis suggests that the bulk of these imports came from the United States.[10]

[9] It is important that the conflict of interests be noted as a potential one. To unequivocally state that this was a conflict in the case of Chile we would need to make the case that the rise in imports of manufactured goods from the United States diminished Chile's ability to industrialise. That is beyond the scope of this work. We also do not have data for Chile's manufacturing value-added as a percentage of GDP prior to 1960. Post 1960, manufacturing value-added grew from a fifth of GDP to a quarter by 1970 (World Bank, 2019b).

[10] A total of 89% of Chilean merchandise imports were from high-income countries in 1960 (World Bank, 2019b, l. 216). There is an 89% correlation between the indexed value of Chilean manufactured imports and US manufactured exports from 1938 to 1960. As Chile's imports increased, the US exports increased. By contrast, the correlation between Chile's manufactured imports and its largest neighbours' manufactured exports Argentina and Brazil for the same period is 11% and 4% respectively (United Nations, 1962, pp. 9, 17, 20, 52).

Besides the potential conflict between investment recipient countries and investor and export-oriented countries, replicating the Tennessee Valley Authority template in the Zambian context would have entirely different consequences for Zambia according to Rostow's theory. Whereas the United States was at the penultimate stage of five stages of Rostow's economic growth with about just a fifth of its work force in agriculture (Rostow, 1960, p. 71; U.S. Census Bureau, 1999, Table 1430), Zambia in 1968 was at the second stage before take-off because close to three-fourths of its population worked in subsistence agriculture (Park, 1968). Where previously electric utilities had targeted one-fifth of households already modernised, the TVA focused on the four-fifths of households that utilities claimed were incapable of modernisation. Targeting hydropower for cheap electricity consumption in the United States propelled high mass consumption, which drove demand across domestic and already established industries and drove rises in employment, wages and gross value-added. Along with contributions from the Great Depression and the Second World War, this gave rise to the famous Kuznets Curve: the share of income of the two lowest quintiles rose from 13.5% in 1929 to 18% after the second world war, while the share of the top quintiles fell (Kuznets, 1955, p. 4). Inequality fell. This was the key difference between the balanced growth targeted by the TVA and the unbalanced growth which ZESCO targeted: the TVA was for those excluded; ZESCO was for those already included. 85% of Zambian electricity consumption was by the export-facing copper mining industry (IBRD and IDA, 1973, p. 1) initiated by colonialists. Zambia's copper was valuable to high-income markets since Zambia was second only to the United States as a non-communist supplier of the commodity.[11] Export of raw metals and ores have continued to account for 70–80% of Zambia's exports (World Bank, 2020f). Targeting hydropower to benefit extractive mining industries in Zambia would benefit Western markets, those employed in mining, those

[11] Prior to independence, mining was dominated by two companies, Anglo American and Roan Selection Trust, who paid royalties to the British South Africa Company in lieu of a local national government. The geographic area that would become to be known as Zambia was the world's third largest producer of copper in the world, trailing only the United States and USSR with 14% of global output (Barton, 2016, pp. 20–21). The Government of Zambia nationalised the mines in 1969 and reprivatised them in 2000 following protracted and failed negotiations for the Government of Zambia starting in 1992 (Barton, 2016, pp. 21, 141–144).

serving mining companies and the customers of the mining products. The beneficiaries within Zambia would be limited, and inequality would rise.

Ironically, Kuznets' observation of reduced inequality from a time-bound experience of one country's experience with a high degree of government intervention targeted at the bottom of the pyramid and hence with balanced growth had subsequently been used to justify development along a capitalist paradigm of unbalanced growth. Unlike Rostow's *Stages of Economic Growth: A Non-Communist Manifesto*, Kuznet's theory that 'income inequality would automatically decrease in advanced phases of capitalist development, regardless of economic policy choices or other differences between countries, until it stabilised at an acceptable level' (Piketty, 2014, p. 13) was not advertised as a political one.[12] But like Rostow's modernisation theory and like the many calls for the 'weaponisation' of the Tennessee Valley Authority, Kuznets' theory was a partisan one designed to further US political interests and ideology in the context of the Cold War (Piketty, 2014, p. 13).

Accompanying Rostow's logic couched, in economic terms, was political pressure from the United States to export its 'best-known, most highly appreciated institution' (Cooke, 1949; Ekbladh, 2002, p. 350)—the Tennessee Valley Authority. The historian Arthur Schlesinger called the Tennessee Valley Authority 'a weapon which, if properly employed, might outbid all the social ruthlessness of the Communists' (Schlesinger, 1949; Ekbladh, 2002, p. 350). Even a member of the judiciary was recorded making overtly political statements about the Tennessee Valley Authority. 'The word "Tennessee" is well known all the way across from the Mediterranean to the Pacific', said US Supreme Court Justice William Douglas in 1951. 'They know about Tennessee because they have heard of the Tennessee Valley Authority [...] that fits their needs and will solve many of their basic problems. The TVA can also be utilised as one of the major influences to turn back the tide of communism which today threatens to engulf Asia' (Douglas, 1951; Ekbladh, 2002, p. 335). The political resilience and popularity of the TVA carried through into the 1960s as US president Lyndon Johnson's talked of replicating the Tennessee Valley Authority in Vietnam in his speech 'Peace Without Conquest', delivered a month after he had dispatched 3500 US Marines to South Vietnam to commence the United States' ground warfare in the Vietnam War:

[12] Stated visually, the Kuznets' Curve was a statement of a universal 'n' shaped curve where the x-axis represented economic growth and the y-axis represented inequality.

'Electrification of the countryside—[...] that [...] is impressive [...] In the countryside where I was born, and where I live, I have seen the night illuminated, and the kitchens warmed, and the homes heated, where once the cheerless night and the ceaseless cold held sway. And all this happened because electricity came to our area along the humming wires of the REA [Rural Electrification Administration].' Vietnam's Mekong River could provide food, water and power, he told his audience, on a scale that would 'dwarf even our own TVA'. (Johnson, 1965)

Yet as the evidence of electrification in Zambia shows as well as investment appraisal documents for hydropower projects across 18 countries listed in Appendix A, Johnson's experience and vision of electrified rural homes from hydropower investments abroad was at odds with the reality of the Bank's focus on energy catering to mining and industry. This happened because proponents of the TVA template within the Bank had forgotten the processes by which the TVA had achieved an increase in employment and stimulated the economy, and focused only on the outcomes.

The Tennessee Valley Authority (TVA) template came to be used as a universal supply-driven solution—to be referred to as the 'granddaddy' of all regional development projects (Scott, 1998) and as 'the marquee example of government-funded economic development' (Staples, 2006, p. 12).[13] By 1951, the Bank had extended 32 loans in 13 developing countries.[14] Energy accounted for 60% of the Bank's development finance portfolio (IBRD, 1951c, pp. 54–56).[15] In 1956, the Bank initiated the Economic Development Institute (Mason & Asher, 1973, p. 326) that would train future finance and prime ministers. The institute took students on outings. The Tennessee Valley Authority's network of dams, power plants, irrigation systems and fertiliser plant proved to be 'particularly popular [...] as many Third World countries [*sic*], as well as the World

[13] Staples continues: '[The TVA] became the model for similar river-development schemes throughout the United States and the world, as well as for the Rural Electrification Administration, which brought electricity to rural areas across the country in order to encourage their economic development and modernisation' (2006, p. 12).

[14] I have counted as developing countries Brazil, Chile, Colombia, El Salvador, Ethiopia, India, Iraq, Mexico, Nicaragua, South Africa, Thailand, Turkey and Uruguay, and excluded Australia, Belgium, Denmark, Finland, France, Iceland, Luxembourg, the Netherlands and Yugoslavia.

[15] Using nominal values for 1948–1951 loan agreements, electric power accounted for 61.5%, transport for 21.8%, agriculture for 9%, foreign exchange for finance for 4.6%, flood control for 2.8% and telecommunications for 0.3%.

Bankers, saw it as a model of the type of large-scale economic development that could be produced by large infusions of outside development capital under close governmental supervision'.

The TVA would allow the World Bank 'the relative ease of transferring the technology of power generation from more developed to less developed countries' (Mason & Asher, 1973, p. 713). That it was very much a supply-driven solution is borne out in early World Bank country reports requesting water and sanitation investments that were rejected, crescendoing into the outcry by developing countries in 1960 that resulted in the creation of the International Development Association (Ahmed, 2020a).[16] The TVA itself was predicated on cheap electricity for all, but in the developing country contexts where it was introduced, it was used to supply those already with electricity access more. The TVA was an adaption of Henry Ford's model of mass production, applied to electrification (Tobey, 1996, pp. 93–96) to stimulate aggregate demand (Barber, 1996, pp. 128–131) and gave birth to the bottom-up Keynesian multiplier effect. As a supply-side solution offered by the World Bank, regard was not paid to how different the conditions were in Zambia versus what they had been in the United States.

2.3.3 *The Bank's Economic and Financial Rationale for Investing in Hydropower*

Having understood the context of the Tennessee Valley Authority, its successes within its context, the broader global political context underscoring its propagation and the creation and use of the World Bank by the United States, as well as Rostow's modernisation theory that married US interests with international development economic theory, we can now zoom back

[16] The International Bank for Reconstruction and Development's Cuba mission reported that were a water crisis allowed to occur, 'all employment in all industries would cease' (IBRD, 1951b, p. 328). The IBRD's Nicaragua report called investments in sanitation, education and public health 'without question' the first priority to combat high disease rates and low nutrition-inhibiting productivity (IBRD, 1953b, p. 22). However, no loans were made to Cuba in the 1950s where requests were made for water projects (Alacevich, 2009, p. 129). Of the 11 loans to Nicaragua between 1951 and 1960, none was for water, sanitation, health or education (Kapur et al., 1997, p. 111). With Paraguay, it was a similar story. Having accepted a US Eximbank loan in 1954 for a water supply system, the Bank 'arguing that the project was an unproductive amenity, and moved as well by its rivalry with the Eximbank, dropped plans for lending to that country, an interruption that lasted until 1962' (Kapur et al., 1997, p. 111).

into the economic and financial rationale used by World Bank professionals—leading economists and executives—to justify investments into energy and in particular hydropower.

Broadly, the thinking was as follows:

(a) Economic infrastructure would benefit the economy and generate taxes to pay for welfare assets (which, in the case of access to electricity in Zambia, it did not).
(b) Social infrastructure was not also economic infrastructure.
(c) In any case, it was not worth attempting to appraise the economic benefits of welfare assets either on an ex-ante or on an ex-post basis.

2.3.3.1 Economic Infrastructure Would Benefit the Economy More Quickly Than Social Infrastructure

While the World Bank today no longer persists with the distinction between economic and social infrastructure (though the OECD does persist (OECD, 2019)), it is important to note the World Bank did make this distinction at the time that it financed most of Zambia's existing energy infrastructure. As late as 1995, the World Bank categorised water supply and sanitation as 'social' investments, separate from 'infrastructure' which were telecommunications, transportation, electric power and other energy (Kapur et al., 1997, p. 6).

The words of World Bank vice presidents lend insights into the rationale as to why the Bank as an institution thought there was a dichotomy between social and economic infrastructure.

In rejecting an investment for urban water in Colombia, World Bank vice president Robert Garner said that the Bank would only consider municipal projects if 'tangible evidence was provided that the projects were closely connected with the development of productive facilities' (Garner, 1952; Alacevich, 2009, p. 114).[17] Rejecting the project once again a year later in 1953 following a resubmission for the application of a loan for the municipality, Garner argued that the 'Bank should concentrate its efforts on projects which will yield the <u>greatest</u> and <u>quickest</u> increase in output and productivity'. Out of these new sources of income

[17]The request accompanied requests for investment in other sectors, highlighting the importance of power: 'Electric power is perhaps the most single element in modern economic development' (Currie, 1950, p. 514).

would come the means for member countries 'provide out of their own resources better municipal services, better housing, better health and education' (Garner, 1953; Alacevich, 2009, p. 117). Garner's sentiment would be reiterated by his successor almost a decade later. As World Bank vice president Burke Knapp (Olivier, 1961, p. 33) explained in 1961,

> We can lean a little in the direction of taking those things which have come to be known as social projects rather than directly productive economic projects. My own definition of these is things that are less investment in future productivity and more satisfaction of current welfare requirements, like housing, water supply, other municipal services. Our sort of doctrine in the main in the past has been that those things were the fruits of economic development and that we would rather invest in the means of economic development and let countries develop the taxable capacity and the productivity that would enable these amenities to be provided.

Arguments that potable water, considered social infrastructure, was key to productivity fell on deaf ears. An IBRD mission to Cuba reported that were a water crisis allowed to occur, 'all employment in all industries would cease' (IBRD, 1951b, p. 328). The Nicaragua report called investments in sanitation, education and public health 'without question' the first priority to combat high disease rates and low nutrition-inhibiting productivity (IBRD, 1953b, p. 22). However, no loans were made to Cuba in the 1950s where requests were made for water projects (Alacevich, 2009, p. 129). Of the 11 loans to Nicaragua between 1951 and 1960, none was for water, sanitation, health or education (Kapur et al., 1997, p. 111). It was a similar story with Paraguay. Having accepted a US Eximbank loan in 1954 for a water supply system, the Bank 'arguing that the project was an unproductive amenity, and moved as well by its rivalry with the Eximbank, dropped plans for lending to that country, an interruption that lasted until 1962' (Kapur et al., 1997, p. 111).

In relation to the rejected water investments, Currie (1979; Alacevich, 2009, p. 129) complained that the Bank 'had the strange belief that water was social, not economic'. The categorisation of water remains as such, with water and sanitation listed under 'Social Infrastructure and Services' and energy listed under 'Economic Infrastructure and Services' on the OECD's website (2019). It was not until a political uprising from developing members that the Bank made its first urban water supply loans in 1960 with the creation of the International Development Association, and even then, investments mainly privileged the urban middle-classes.

But lower marginal productivity could not be the reasoning for denying investment in water in Cuba and Nicaragua where clean water was seen as a prerequisite for productivity and industry. In truth, Garner's opposition to investing in water was informed by lived experience and emotive prejudice. Flippantly, in an interview in 1985, he said: 'Why should such and such a country have a water supply system in its town? When I was brought up in Mississippi [...] we didn't have any water in our house' (Kapur et al., 1997, p. 110). Kapur, Lewis and Webb remark that 'one is left wondering [...] had Garner been born in a town with piped water in each home, he might have agreed to lend for municipal water services' (1997, p. 125). Clearly in addition to *Erfahrungwissenschaft* at play among the economic theories used to justify investment in power, *Erlebeniswissenschaft* was also at play in the preclusion of social infrastructure.[18]

Among the *Erfahrungwissenschaftler* at the World Bank were Paul Rosenstein-Rodan, for whom the TVA power area's growth provided a template for the importance of complementary knock-on effects between sectors of the economy, and Albert Hirschman, who propounded unbalanced growth. Both were economists who served at the World Bank.[19] Rosenstein-Rodan argued for a balanced growth approach (1943, pp. 205–206) illustrating with a toy example of a shoe factory employing 20,000 formerly agrarian workers in a more-or-less closed and self-sufficient domestic economy would be useless. The employees would require more than the shoes they were able to buy with their wages:

> If, instead, one million unemployed workers were taken from the land and put, not into one industry, but into a whole series of industries which produce the bulk of the goods on which the workers would spend their wages, what was not true in the case of one shoe factory would become true in the case of a whole system of industries: it would create its own additional market [...] The industries producing the bulk of the wage goods can therefore be said to be complementary.

By contrast, Hirschman argued that developing countries had an insufficient endowment of resources to enable them to invest simultaneously in all sectors to achieve balanced growth and take-off (Hirschman, 1959;

[18] *Erlebeniswissenschaft* is knowledge acquired from lived experience, as opposed to *Erfahrungwissenschaft* which is based more on observations (Fukuyama, 2018).

[19] Paul Rosenstein-Rodan, a University College London economist, became the IBRD's Economic Department Assistant Director. Albert Hirschman also consulted for the IBRD.

Ncube et al., 2017, p. 3). According to his unbalanced theory of growth, therefore, governments had to invest strategically in infrastructure to raise the productivity of selected industries, not to the broad extent that Rosenstein-Rodan had imagined, to multiply new investment opportunities (Hirschman, 1959; Ncube et al., 2017). Availability of reliable power and transportation facilities were therefore prerequisites for economic development (Hirschman, 1959; Ncube et al., 2017). The argument for focus on selected industries seems to have prevailed in the US Department of Commerce's analysis of Zambia in 1968 (Park, 1968, p. vii): 'Zambia's bright prospects for continuing economic development arise largely from the country's strong financial base, consisting principally of revenues derived from abundant mineral resources.'

To preserve the IBRD's triple-A credit rating, the Bank established a separate International Development Association (IDA) in 1960 in response to the threat that frustrated governments of low-income countries would move their membership to a rival multilateral bank that would fund social infrastructure projects, and which would be housed by the UN, where each nation had one vote, undercutting the United States' influence. The funding of social infrastructure projects could have been an opportunity to test the hypothesis that so-called social infrastructure projects did not yield the same productivity returns as so-called economic infrastructure, but even as late as 1967, Hirschman argued against ex-post evaluations to rank infrastructure projects, citing the impossibility of using a single scale to amalgamate all the varied dimensions. And so it was that the IBRD continued to favour economic infrastructure projects without validating its assumptions.

This also meant that the IBRD could continue providing loans to sectors with which it was comfortable: hydropower helped it achieve a triple-A credit rating and allowed it to sell bonds on the private market (Sharma, 2017, p. 15). The Bank's credit rating affected its ability to raise funds. Its Wall Street leadership & project lending won it AAA credit rating by the mid-1950s (Sharma, 2017, p. 15), which reinforced for management that its conservative approach of sticking to lending what it knew was working. 'The first loans granted by the Bank [...] proved to be successful, and in fact they were completely repaid', notes Alacevich. 'According to those who urged and considered it inappropriate for the Bank to venture into higher-risk loans in an environment where it was difficult to raise funds, there was no reason to change policy. The Bank took pride in reorganising power companies to which it loaned so that they ran on sound financial lines (Mason & Asher, 1973, p. 716). The financial success of the first loans validated its position against alternative models of "impact" and "social loans" and contributed to perpetuating the model' (Alacevich, 2009, p. 136); the bank had not as of 1971 suffered a financial loss (Mason & Asher, 1973, p. 258) (Table 2.3).

The Bank was thus able to continue using an appraisal methodology developed in the mid-1950s that assessed the need for power using market rather than shadow economic demand, in using market prices, and weighing the case for power in financial rather than economic terms (Mason & Asher, 1973, p. 716). This, then, explains why the Bank's investment appraisal for the Kafue hydropower project in 1973 assessed the electricity demand for Zambia's copper industry (IBRD and IDA, 1973, p. 1) rather than by Zambians. It also explains why industrial, mining or mineral processing demand and not residential demand was the Bank's primary target for its first hydropower investments in Brazil (IBRD, 1950c, 1953a), Colombia (IBRD, 1950a, 1950b, 1951a; World Bank, 2020b), Ethiopia (IBRD, 1961b; World Bank, 2020c), Ghana (IBRD, 1961a, p. 2), Kenya (IBRD and IDA, 1971, p. 3) and Turkey (IBRD, 1952b; World Bank, 2020d). Investment in hydropower in Zambia was representative of the World Bank's institutional bias towards this type of development intervention. Where there were resources for hydropower, the World Bank went. It even went where it could not legally meet demand: the then-chairman of the British construction company selected to build the Kariba Dam recalls that the Bank 'twisted' the Zambian government's arm to make it accept the Kariba North Bank power station, to assist the Bank in circumventing investment sanctions on Southern Rhodesia (Morrell, 1987,

Table 2.3 The IBRD's historic credit rating

Year	*Bank credit rating*
1947	AA—Fitch
	A—S&P
1950	A—Moody's
	AA—Fitch
	A1—S&P
'Soon after 1950'	AA—Moody's
'Since the mid-1950s'	AAA—Moody's, Fitch, S&P

Author's creation. Source: Mason and Asher (1973, p. 132), Ahmed (2020a)

pp. 152–153), referred to in the Bank's earlier investment appraisal documents as 'the Colony' (IBRD, 1952a).

Having understood the theoretical and empirical linkages between energy and industrialisation and the Bank's logic behind its significant support for the expansion of power generation in Zambia, we will now assess how power generation's target beneficiaries mining and industry fared, and how effectively investment into energy provision for them translated into the sort of welfare outcomes that World Bank executives Robert Garner and Burke Knapp suggested that they would.

2.4 A Brief History of Zambia's Industrialisation Until 2015

This chapter has highlighted literature that shows that industrialisation has so far been a reliable path for economic development, and revealed the underlying assumptions of World Bank executives that the tax revenues earned from industry should pay for welfare amenities. This section now reviews how successfully investments into Zambia's power generation have resulted in industrialisation, and what that has meant for the welfare of Zambians in general. It also looks at other factors that have impacted Zambia's industrialisation, namely structural adjustment advocated for by the Bank and IMF, which saw de-industrialisation from 1991 to 2002.

Theory discussed at the beginning of this chapter would predict that investment into energy provision in Zambia for the purpose of industrialisation should have resulted in structural shifts in terms of sectoral employment from low-productivity agriculture to higher-productivity manufacturing, and an increase in the proportion of GDP generated by manufacturing. These shifts should have made the workforce more productive, and with manufacturing's backward and forward linkages and through the multiplier effect enabled by a more prosperous workforce, should generate further demand for goods and services. As these sectors grow and thrive, tax revenues for welfare services should increase. Measures of success that improvements in welfare are being distributed equitably would then be an increase in average expected life expectancy, increase in access to electricity, and increase in GDP per capita. Remembering that the World Bank's significant investments into power generation in Zambia were approved in 1956 and 1970–1974 (Table 2.1), as well in 1964 for a smaller investment, we can assess whether Zambia's power generation achieved either of these predicted outputs or outcomes, before looking at other dynamics that affected Zambia's industrialisation.

2.4.1 An Overview

Looking first at the World Bank and Energy Regulation Board's statistics to assess the relationship between energy consumption and manufacturing value-added and the relationship between manufacturing value-added and GDP we find that the energy consumption and manufacturing value-added are 94% correlated over the period 1971–2018 and that manufacturing value-added and GDP are 98% correlated over the same period. Over the observed period, 1% increase in energy consumption was associated with a 1.74% increase in manufacturing value-added, a statistically significant result with an 80% R-squared (Table 2.4).[20]

Looking second at structural changes in the economy, the broad trends do not neatly follow predictions: manufacturing's contribution to GDP increased eight times between 1960 and 1990, but the proportion of the population employed in agriculture barely changed. Manufacturing

Table 2.4 Regressing the log of manufacturing value-added on the log of total-electricity consumption in the economy

	(1)
VARIABLES	*log_manuf*
log_energy	1.741***
	(0.126)
log_manuf	
Constant	6.782***
	(1.115)
Observations	48
R-squared	0.806

Analysis: Own

Standard errors in parentheses: ***$p < 0.01$; **$p < 0.05$; *$p < 0.1$

[20] The World Bank offers manufacturing value-added expressed in constant local currency (World Bank, 2020f). It also offers energy consumption per capita and total population statistics which enables us to compute results up to 2013. For 2014–2018, we used total energy consumption statistics offered by the Energy Regulation Board as outlined in Table 1. To calculate the effect of 1% changes in the explanatory variable on the dependent variable, I ran log-log regressions.

value-added's contribution to GDP did increase substantially in the period following the Bank's investments into power: from a base of 4% in 1960 to 10% in 1970 to 17% in 1980 and up to 32% in 1992 (World Bank, 2020f). It fell dramatically to 9% in 1994 (World Bank, 2020f) following structural adjustment, and has hovered at approximately that level since. On the other hand, investment in power generation did not apparently have the predicted effect on employment in agriculture. 75% of Zambia's population was assessed as being dependent on subsistence agriculture in 1968 (Park, 1968, p. 6). The sector accounted for 11% of GDP (Park, 1968, p. 6). The next datapoint available for employment in agriculture is 72% in 1991 (World Bank, 2020f). In the time in between, the World Bank invested USD 1.13 billion (Table 2.1) in the early 1970s in Zambia's power generation. The investment seems to have had minimal impact on the structure of employment in Zambia. The decade since 2008, employment in agriculture fell by a quarter to 54% by 2018 (Table 2.1). This cannot be ascribed to investment in power generation 40 years earlier. Likewise, investment in power generation did not have the predicted effect on agricultural contribution to GDP which hovered between 10% and 16% from 1960 to 1988 (World Bank, 2020f). It spiked to 21% in 1992 and 31% in 1992 (World Bank, 2020f) following structural adjustment (more on this later) before falling steadily to 3% in 2018 (World Bank, 2020f).

The conflicting results for manufacturing and agriculture suggest that as Zambia's population grew, employment opportunities in Zambia's small manufacturing sector did not keep up. Quality of life would have improved for urban dwellers able to avail opportunities, but not for the rest of the rural population or the urban poor. This is borne out by the election of Michael Sata as president in 2011 who ended two decades of the incumbent Movement for Multiparty Democracy's rule that had overseen rapid growth in the 2000s. He overcame not only the incumbent, but also eight other opposition parties by tapping into the urban poor's disgruntlement with rising inequality (Resnick & Thurlow, 2014). It is also borne out in several sets of metrics measuring welfare. Zambia's GINI index of inequality has been very high—0.61 in 1991 during structural adjustment (World Bank, 2020f) and possibly 0.66 in 2006 (2014). Inflation-adjusted fell to two-thirds of its value by the end of Zambia's first president's presidency from the time of independence in 1964 over 27 years before rising to its present-day level, which is just 8% above its independence day level, compared to real GDP growth of 450% since independence (World Bank, 2020f).[21] Despite being categorised as a middle-income country by the World Bank, 45% of the population lived on up to USD (2011 PPP) 3.20 per day in 2017 (World Bank, 2020f). Zambia's grid electricity access rate was just 33% as of 2017, with rural grid electricity access at just 6%, in contrast to 66% for urban access (ZICTA and CSO, 2018). 27% of households had no type of electricity at all (ZICTA and CSO, 2018). Just 18% of households had internet access (ZICTA and CSO, 2018).

2.4.2 *A Chronological View*

Zambia's manufacturing sector grew quickly from a small base to service the mining sector from independence in 1964 until 1973 following a policy of import substitution and high copper prices (Gondwe & Pamu, 2014, p. 12), as well as following investments into Zambia's energy provision. Growth slowed over the next few decades as it was constrained by lack of infusion of new technology, the high cost of borrowing and foreign competition (Gondwe & Pamu, 2014, p. 12). The dip in the sector's value-added in the 1990s followed structural adjustment (Fig. 2.3).

Post-independence Zambia was characterised by President Kenneth Kaunda's state-led 'humanist development' (Funder et al., 2018, p. 34). Agriculture and state-subsidised manufacturing were responsible for much of Zambia's growth (Thurlow & Wobst, 2007, pp. 221–222); manufacturing growth has a 97% correlation with the growth of Zambia's GDP (World Bank, 2020f).[22]

However, on the back of rising debt and food prices which caused food riots in the wake of droughts—such that President Michael Sata later

[21] This is not the perfect proxy for the income of the average member of society because on the one hand, it is skewed towards greater income by being calculated on a mean average rather than median average basis and thus overstates average income. On the other hand, inflation-adjusted prices which form the basis of GDP calculations fail to account for the reducing price of technology over time as technologies become cheaper. Tracked over time, however, it still gives an idea of inequality when compared against GDP growth

[22] According to the World Bank, GDP is calculated without making deductions for depreciation of fabricated assets or for depletion and degradation of natural resources and is calculated as the sum of gross value-added by all resident producers in the economy plus any product taxes and minus any subsidies not included in the value of the products (World Bank, 2020f). The Central Statistical Office compiles Zambia's GDP statistics and has since 1994 been including the informal sector, which it caveats may exaggerate the total GDP figure relative to other countries (Jerven, 2013, p. 48).

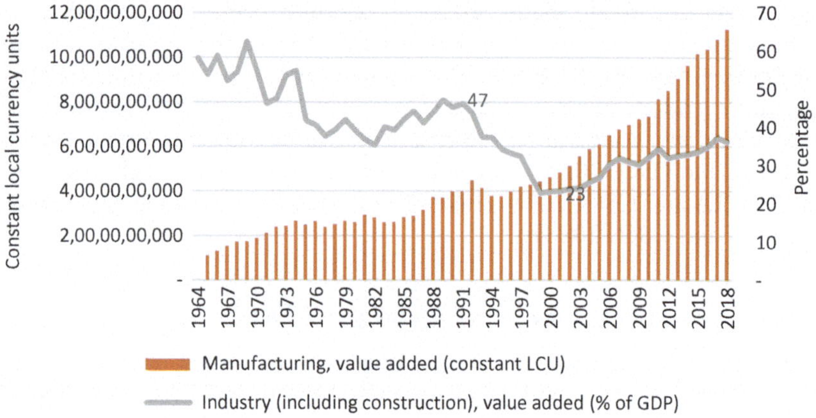

Fig. 2.3 Zambia's manufacturing value-added has generally been growing. (Data source: World Bank, 2020; graph: own. Industry corresponds to international sector industrial classification (ISIC) divisions 10–45, whereas manufacturing is a subsector, divisions ISIC 15–37 (World Bank, 2020a). ISICs 41–43 are construction, ISICS 36–39 are water supply, sewerage, waste management and remediation, p. 43. However, the UN and this study counts ISICs 10–14 as manufacturing also: ISIC 10 is manufacture of good products, 11 is of beverages, 12 of tobacco products, 13 of textiles and 14 of wearing apparel (UN, 2008, pp. 87–101). ISICs 10–12 constitute Zambia's largest manufacturing subsectors and ISICs 13–14 were important subsectors until 1991)

remarked in 2013 that 'Kaunda's government was overthrown because of food riots and I don't want food riots'—multiparty elections were run in 1991 (Reuters, 2013; Funder et al., 2018), ushering in a change of regime and political economic philosophy. Moving away from donor dependency and the old regime's emphasis on state-led development, the new government of Movement for Multiparty Democracy government adopted the Washington Consensus advocated by the Bretton Woods organisations (the International Monetary Fund and the World Bank), which was

> predicated on the belief that by eliminating 'distortions' in the economy, Africa would grow faster by constructing an economy based on principles of free and unfettered markets. In reality, these structural adjustment policies foisted on developing countries have actually discouraged industrial development. (Stiglitz, 2017, p. 12)

Per the Washington Consensus, the Movement for Multiparty Democracy government followed a programme of structural adjustment, which it largely achieved by 1998. This included privatisation of state owned enterprises and trade liberalisation by reducing trade protection, which left Zambia one of the most open economies in Africa (Thurlow & Wobst, 2007).

As the data across metrics above suggest, structural adjustment in 1991 wreaked havoc across the economy. Import tariffs fell by two-thirds on imported textiles, leading to increased cheap imports from Asia, against which Zambia's textiles subsector was unable to compete (Resnick & Thurlow, 2014). Compounding the problem was the import of second-hand clothes donated to foreign charities and sold at low prices (Resnick & Thurlow, 2014). Employment in textiles fell from 250,000 in the 1980s to less than 2500 in 2002, and its contribution fell from 1.8% of GDP in 1991 to less than 0.3% in 2010 (Resnick & Thurlow, 2014). Manufacturing value-added's proportional contribution to GDP fell to a third of what it had been. Value-added per worker fell by 20% from 1991 to 2002, almost entirely due to negative structural change, as manufacturing job losses resulted in people seeking survival in low-skilled subsistence farming and in informal trading, resulting in a reduction of national labour productivity (Wobst & Thurlow, 2005; McMillan & Rodrik, 2011; Resnick & Thurlow, 2014; Mulanda & Punt, 2020). Life expectancy fell to its lowest since Zambia's independence, although the causality in increased inequality and mortality is confounded by an increase in the prevalence of HIV at the same time (World Bank, 2020f), as well as drought (Beilfuss, 2012).[23]

A couple of authors have blamed Dutch disease (the adverse impact on other sectors arising from the export of natural resources) for the decline of textiles, but without comprehensive analysis. Resnick and Thurlow state that compounding the textiles subsector's woes was high copper prices, making Zambia's real exchange rate appreciate and thus making imports cheaper relative to domestic production (Resnick & Thurlow, 2014, p. 9). Where according to World Bank statistics,

[23] Life expectancy in 1960 was 45, increased to 49 by 1970 and then more or less plateaued at 51 from 1974 to 1982 (World Bank, 2020f) before steadily declining. It reached a nadir of 43 between 1993 and 1997 (World Bank, 2020f) following structural adjustment, but the timing may just have been a coincidence: life expectancy began its decline nine years earlier. The World Bank started recording the incidence of HIV for Zambia in 1990 when it was 2.4%, peaking in 1992 at 2.6% before declining. The coefficient of correlation between HIV incidence and life expectancy between 1990 and 2018 was –80%, with life expectancy reaching 62 years in 2018 (World Bank, 2020f)).

mineral rents in 1970 accounted for 35% of GDP, in 2017, they accounted for 12.5% (World Bank, 2019a). Nevertheless, Zambia's export value still follows the price of copper, the most commonly mined mineral in Zambia (see Fig. 2.4); copper and copper products accounted for 72% of Zambia's exports (World Bank, 2020e).

Cardozo et al. also suggest that with higher copper revenues resulting in the appreciation of the Kwacha, textile imports from China became more competitive (Cardozo et al., 2014, p. 7) resulting in the shrinkage of Zambia's textile manufacturing subsector (Cardozo et al., 2014, p. 10). There are weaknesses with how both sets of authors present the argument. With Resnick and Thurlow, they mention it as a confounding factor, but it may as well be a coincidental factor. The evidence Cardazo et al. present is the fact that the textiles sector shrank. However, all of Zambia's other manufacturing subsectors grew over the same time period (Cardozo et al., 2014, p. 10). Looking to the data, when one divides World Bank historic data for GDP per capita in current local currency by GDP per capita in current US dollars (World Bank, 2020f), one gets average rates of exchange between the Zambian Kwacha and US dollar. If one uses 2000 or 2001 as

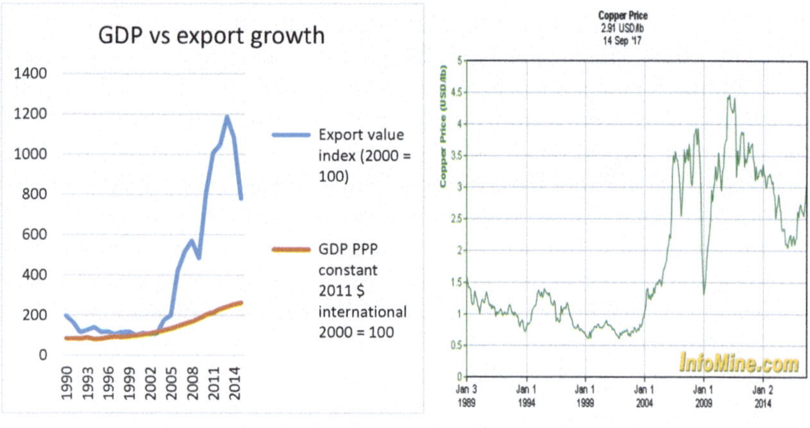

Fig. 2.4 Historic copper prices, GDP and export growth. Copper prices have been driving Zambia's overall export and GDP growth, in spite of the percentage of non-mining companies that export decreasing. (Data sources for graphs: World Bank, 2017c (author's creation); Glacier Resource Innovation Group, 2018)

the base year for looking at whether the Kwacha appreciated or depreciated over the 2000s, in no year did it appreciate beyond the base year. In 2006 it returned to the exchange rate at which it was in 2001. In that year, the Kwacha did appreciate by 19%, but then depreciated by 11% the following year and by 35% in 2009. Mining may well have been the cause of currency fluctuations in 2006, but the link is not definitive. Figure 2.4 shows that global copper prices rose exponentially from 2006 to 2008, and that Zambia's export value continued to grow beyond then, but the annual average Kwacha-dollar exchange rate remained within a band of 3.6–4 Kwacha per dollar.

If 2006 was indeed the year in which Dutch disease harmed textiles, 2006 seems to be an aberration. Gondwe and Pamu argue that thanks to generous incentives and subsidies to the mining sector in Zambia, mining companies have little incentive to convert foreign exchange earnings into Kwacha to settle domestic obligations and that as a result, insufficient mineral rent gets absorbed into the domestic economy (Gondwe & Pamu, 2014). The depreciating trend of the Kwacha relative to the dollar suggests that Gondwe and Pamu are at least generally correct. In a case of Dutch disease, the Kwacha would appreciate. Instead, the Kwacha had half the US dollar value in 2018 that it had in 2012 (World Bank, 2020f).[24] Moreover, while metal and ores have accounted for between 70% and 80% of the country's exports since 2005, manufacturing exports have increased from 9% to 14% (World Bank, 2020f). Positive spill-over effects would be greater if this was the other way round since manufacturing has greater forward and backward linkages to the rest of Zambia's economy than mining does, which also has limited potential for direct employment and historically has not generated much tax revenue (Resnick & Thurlow, 2014, p. 8).

[24] The only other relevant paper in the first 20 recommended articles that the search item 'Dutch disease Zambia' yielded in UCL's online library was Boos and Holm-Müller's paper (Boos & Holm-Müller, 2016). The paper makes the case that patterns of Genuine Savings—the reinvestment of rents from the depletion of natural capital rents into physical or human capital—can be used to indicate whether a country is experiencing the 'resource curse' of Dutch disease (Boos & Holm-Müller, 2016). However, Boos and Holm-Müller overlook an alternative reason to that of the resource curse for why Genuine Savings do not materialise—the one pointed out by Gondwe and Pamu that substantial rents were not collected by the economy in the first place. Their analysis that the lack of Genuine Savings is indicative of a resource curse cannot therefore be sustained.

But for textiles, the 2000s saw a general rebound in Zambia's economic growth which Bhorat et al. (2017) attribute in large part to a rebound in copper production, Bayliss and Pollen (2019) attribute to a boom in commodity prices, as well as favourable weather that supported agricultural production, and sustained macroeconomic stability,[25] and Barton attributes to greater institutional transparency in government and credible government commitment to investors (Barton, 2016). Resnick and Thurlow characterise the 2000s as a period of rapid economic recovery, marking a return to pre-crisis levels and middle-income status (Resnick & Thurlow, 2014). Value-added per worker increased by 3.6% per year from 2002 to 2010: about half of this increase was due to positive structural change driven by faster employment growth in services and a relative decline in farm employment (Resnick & Thurlow, 2014, p. 4).

During the 2000s, the largest contributors to growth were construction, transport, communication, finance and business services which collectively accounted for 56% of the increase in GDP (Resnick & Thurlow, 2014, p. 12). But because these sectors were driven mainly by rising value-added per worker rather than an increase in the number of workers—for example communications was capital intensive, while financial services required highly skilled workers—they generated only 14% growth in employment (Resnick & Thurlow, 2014, p. 12). From 2002 to 2010, manufacturing accounted for 4.5% of increase in national GDP per capita and 1.3% of employment growth (Resnick & Thurlow, 2014, p. 9) with food processing accounting for 70% of the manufacturing sector's contribution to GDP, up from 50% in 1991 (Resnick & Thurlow, 2014, p. 9). Vertical integration with South-Africa based supermarket chain helped Zambia's largest meat processor Zambeef expand (Resnick & Thurlow, 2014, p. 9). Wheat millers benefitted from liberalisation and an appreciating local currency because imported wheat was cheaper and expanded their production of flour and bread for urban consumers (Resnick & Thurlow, 2014). Food-processing was less labour intensive than the declining textiles subsector and contributed less to national job creation than it did to economic growth (Resnick & Thurlow, 2014). Overall, the sector continued to grow. Over 2002–2015, Zambia's real manufacturing value-added grew at a compounded annual growth rate of 5.4%.

[25] Wamulume Kalabo, Chairman of the Zambia Chamber of Commerce and Industry (ZACCI) from 2004 to 2007 attributes growth in the 2000s to better governance (Kalabo, personal communication 4 June 2018; Ahmed, 2021).

Zambia started seeing greater institutional transparency with the presidency of Levy Mwanawasa, who was popularly credited with fighting corruption (BBC, 2008), and with his successor Rupiah Banda. Figure 2.5 shows that Zambia's corruption index score by Transparency International was 26 when Mwanawasa entered office in 2002 but improved to a score of 30 in 2009, months after he had left in September 2008, and continued improving until 2015. Barton has also characterised President Mwanawasa's reforms as reinforcing the Zambian state's 'credible commitment' to supporting a more inclusive institutional environment which helped lower investors' perceptions of future policy uncertainty (Barton, 2016, p. 157).[26] Net inflows of foreign direct investment picked up substantially with the passing of the Zambia Development Agency Act in 2006 and the creation of Zambia's first multi-facility economic zone in 2007 (Barton, 2016, pp. 157, 160) (see Fig. 2.7). Like legislation passed in 1992 which only elicited a minor spike in growth of foreign direct investment (see Fig. 2.7), the Zambia Development Agency Act prohibited expropriation of private property without prompt compensation of the property's fair market value less taxes, levies and duties, but it also offered tax incentives for new businesses, import duty waivers on raw materials and capital goods and deferment of VAT on machinery and equipment (Barton, 2016, pp. 158–163). Coupled with Zambia's first MFEZ, the Act that signalled the Zambian government's credible commitment. By 2008 following the

[26] New institutional economist and Nobel Prize winner Douglass North and political scientist Barry Weingast pioneered this concept in assessing the five significant institutional changes that followed the triumph of the English parliament over the monarchy in the Glorious Revolution of 1688/89 (Coffman et al., 2013). The institutional changes resulted in the 'credible commitment' that the government would not default on its future debt. The changes they observed removed the monarchy's underlying sources of expediency, increased parliament's agency and monitoring, introduced the credible threat of dethroning the sovereign and created a balance between parliament and the monarchy (North & Weingast, 1989; Coffman & Neal, 2013). Building upon this scholarship, modern commentators have added in the British case parliamentary supremacy over public finance, transparency in public accounting, accountability via direct creditor action and robust secondary markets, a stated commitment to maintaining the public faith, the use of the common-law procedure to adjudicate disputes between taxpayers and the regime as well as the perception of stability and permanence of a given regime (Coffman, 2013). Beyond Zambia's structural change, the ZDA Act and signalling of credible commitment with the creation of MFEZs, new institutional political economists espousing the 'credible commitment' theory would observe the parallel between England with a strengthened parliament following the deposition of the monarch James II and Zambia's second decade of multiparty democracy following Kenneth Kaunda's 27-year-long presidency.

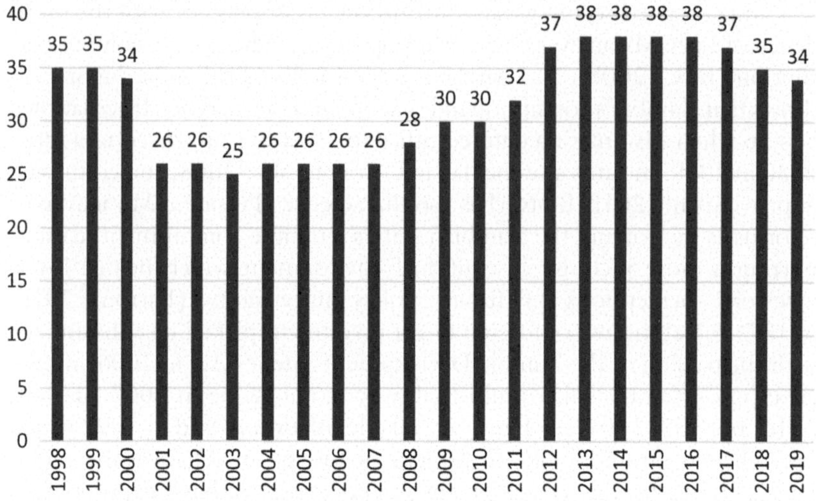

Fig. 2.5 Transparency International Corruption Perceptions Index score out of 100 for perceived corruption within Zambia's public sector (100 = clean, 0 = highly corrupt). (Source: Transparency International, 2020)

announcement of Zambia's fourth MFEZ, the MFEZ initiative had attracted more than USD 3 billion of new investment pledges (Barton, 2016, p. 164). Investors saw goodwill on the president's part, and that he would not change policy without consulting the private sector (Kalabo, personal communication, June 4, 2018; Ahmed 2021).[27] Consequently, manufacturing firms such as Trade Kings grew substantially under his presidency, and so too did the demand for energy (Kalabo, personal communication, June 4, 2018; Ahmed, 2021) (Figs. 2.5, 2.6, and 2.7).

In spite of the growth overseen in the 2000s by the Movement for Multiparty Democracy under presidents Levy Mwanawasa and Rupiah Banda, in 2011, 43% of Zambians voted into power a new president from a new party, defeating the incumbents as well as eight other opposition parties that could have split the vote a lot more than they did. Michael Sata of the Patriotic Front came into power by tapping into the disgruntlement of the urban poor, highlighting their exclusion from the economic recovery (Resnick & Thurlow, 2014). Notably in rural areas, most of which he

[27] Wamulume Kalabo, Chairman of the Zambia Chamber of Commerce and Industry (ZACCI) from 2004 to 2007.

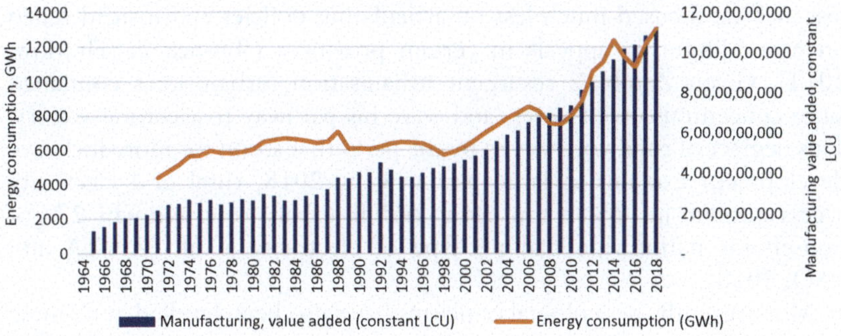

Fig. 2.6 There is a 94% correlation between Zambia's energy consumption and manufacturing's value-added. (Data sources: World Bank, 2020)

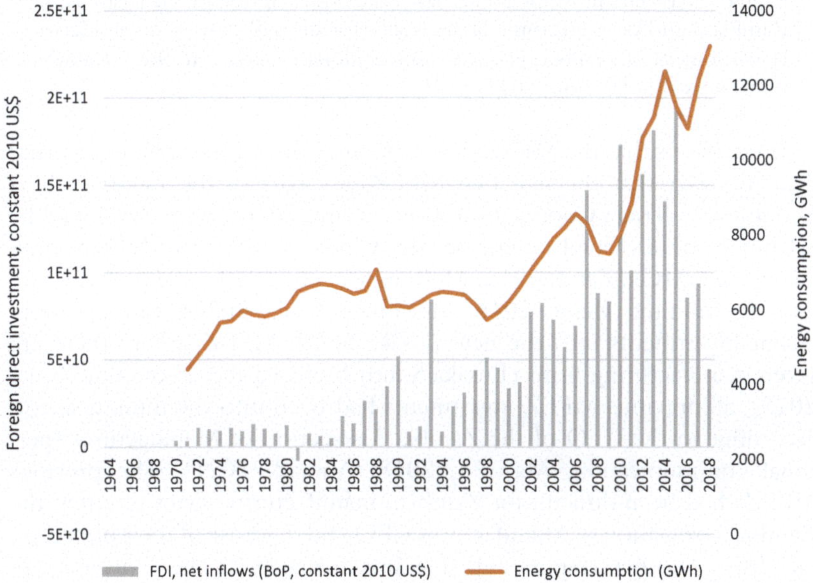

Fig. 2.7 Net inflows of foreign direct investment fell after the power outages of 2015. (Data sources: World Bank, 2020; Energy Regulation Board of Zambia, 2016, p. 9, 2017, p. 9, 2018, p. 36, 2019, p. 39)

lost in,[28] he focused much less on articulating policies and instead relied on ethno-linguistic appeals in certain provinces (Resnick & Thurlow, 2014). Given Zambia's resurgent urbanisation, urban areas contained large concentrations of votes, and were his pathway to electoral victory. This neglect of rural voters even by the party that stood for more inclusive development explains in large part why by 2018, rural grid electricity access stood at just 6% in contrast to 66% in urban areas, and why 27% of households nationwide had no type of electricity at all (ZICTA and CSO, 2018).

More recently, neocolonial criticisms have also been levelled at Chinese investments into Zambian multi-facility economic zones (MFEZs). Ostensibly designed to industrialise the economy and diversify away from mining,

> China's involvement in the zones has come under significant criticism from Zambians and Western commentators alike for allegedly having manipulated [Government of Zambia] policy to gain exploitative access to the country's mineral wealth. (Barton, 2014, p. 91)

Impressions that the MFEZs favour Chinese investors could be inferred from the fact that the first two MFEZs are part of the Zambia-China Economic & Trade Cooperation Zone (Zeng, 2016), were developed by a Chinese state owned enterprise, the China Nonferrous Metal Mining Group (CNMC) (Zeng, 2016; McKinney et al., 2020), and that they are used for mining (Zeng, 2016; McKinney et al., 2020). But for more recent investments into the new Lusaka South MFEZ, where there are seven manufacturing firms (Lusaka South Multi Facility Economic Zone, 2018), all previous MFEZ investments had been into the mining sector, according to the CEO of the Zambia Association of Manufactures (personal communication, May 28, 2018; Ahmed, 2021). The previous MFEZs had been difficult for Zambian manufacturing firms to enter, the Zambia Association of Manufactures CEO said, because of their high capital costs. Another respondent, a grants manager at the University of Zambia, felt that Chinese direct investment was crowding out access to capital for indigenous businesses and government policies were not geared at nurturing Zambian businesses (personal communication, May 29, 2018; Ahmed, 2021). In spite of Barton's impression that MFEZs created

[28] Sata won two rural constituencies (Resnick & Thurlow, 2014).

more institutional inclusivity, a third respondent, a retired director at the Central Bank of Zambia, observes that foreign direct investment has come only from China and suspects that Chinese investment and construction contracts are secured with corruption (personal communication, May 28, 2018; Ahmed, 2021). Chinese construction projects costs, he asserts, are inflated and commit the government to high debt levels. Whereas the respondents I interviewed believed that Chinese state enterprises had developed a competitive edge over indigenous businesses by ignoring good governance, Barton subscribes to the 'wide open door' model that China has merely inserted itself into the neoliberal arrangements that the West opened up because, he argues, Chinese investment facilitates 'the [Government of Zambia]'s longer-term policy to align its economy with the dominant neo-liberal world order' (Barton, 2014, p. 91).

Whether one wants to characterise foreign direct investment in Zambia as neocolonial or neoliberal, it should not be surprising that while Zambia did see economic growth as a result of the investment and the accompanying infrastructure to facilitate it, the economic growth was highly concentrated in the hands of the few who had inherited control of the country's resources, and that its GINI coefficient by 2015 would be a highly unequal 57 (World Bank, 2020f).[29] While life expectancy improved substantially after hitting a nadir in the mid-1990s following World Bank and IMF-informed structural adjustment (Barton, 2016; Stiglitz, 2017) combined with a drop in copper prices (MacroTrends, 2020), the rise in life expectancy since the mid-1990s cannot be directly attributed to the World Bank's investments 20 years earlier. If anything, the investments taken together with the structural adjustment advocated by Bretton Woods organisations in the 1990s perpetuated a colonial legacy of Zambian economic dependence on copper and lack of resilience to fluctuation in raw material prices.

[29] Stuart Barton characterises the presidency of Zambia's first president, spanning from independence in 1964 to 1991, as highly exclusionary both politically and economically (Barton, 2016, pp. 66–90). Real GDP per capita fell by 33% while real GDP had grown by 59% and life expectancy remained stagnant at 48 years of age (World Bank, 2020f). The UNDP notes that the flow into the state's budget from copper revenues has not been very different under either state or private ownership (Simpasa et al., 2013, p. 18).. Weak revenue generation under state ownership is attributed to poor performance of the sector (Simpasa et al., 2013, p. 18). Weak revenue generation under private ownership is attributed to contracts skewed in favour of the concessionaires, as well as illicit financial flows and transfer pricing schemes (Simpasa et al., 2013, p. 18).

* * *

This chapter showed the salience of manufacturing in theory to economic development, the salience of cheap energy supply to economic recovery and industrialisation in the context of the Tennessee Valley Authority; the export of the Tennessee Valley Authority template by the World Bank to low-income countries with very different employment structures and the underlying politics and economic rationale for doing so; and how the links between energy supply, industrialisation and economic development played out in the Zambian context, where structural adjustment resulted in de-industrialisation, deurbanisation and increased inequality in the 1990s before a resurgence in industrialisation but continued inequality and voter discontent.

Almost half a century after the World Bank's investments in Zambian energy, only a third of Zambians had access to grid electricity (ZICTA and CSO, 2018). Because the World Bank invested in hydropower in Zambia when the initial conditions did not match those when the Tennessee Valley Authority was established, the results were also different. The Tennessee Valley Authority was built in the 1930s for those whom had been excluded. It democratised electricity and electrical appliances, stimulated the economy in a balanced manner from the bottom-up and contributed to a more equal society. By the mid-1950s, however, the World Bank's focus on its creditworthiness resulted in an appraisal methodology that assessed the need for power using markets rather than shadow economic demand, in using market prices, and weighing the case for power in financial rather than economic terms (Mason & Asher, 1973, p. 716). When the World Bank came to finance an expansion of ZESCO's energy generation capacity, it came to serve those already included. Unbalanced growth would be an understatement of what it promoted. It promoted an extremely stilted and highly unequal growth.

Rostow's own simple heuristics predicted a difference in outcomes: the United States was at the penultimate stage of five stages of Rostow's economic growth, whereas Zambia in 1968 was at the second stage because close to three-fourths of its population worked in subsistence agriculture. But the differences in outcome also shows which argument out of balanced versus unbalanced growth ended up being more correct. Whereas the Tennessee Valley Authority fuelled construction projects executed by domestic firms, fed into agriculture and tied up with other New Deal programmes that also promoted rural electrification to deliver balanced growth, foreign firms constructed Zambia's hydropower dams,[30] whose primary beneficiary was extractive export-forcing industries with few linkages to the domestic economy.[31] Broader electrification programmes for residential consumption were evidently not promoted so that even until today, two-thirds of the nation is without grid access. Power supply for industrialisation, on the other hand, does seem to have been successful: industrialisation grew rapidly following investment into power and expansion of Zambia's mining capacity. This industrialisation however was undone by the Bank and IMF's structural adjustment policies implemented by the Movement for Multiparty Democracy so that even today, upwards of 70% of the nation's exports are metals and ores. Unbalanced growth in Zambia failed to deliver an eventual balancing. Having interrogated the difference in context between the World Bank's template for hydropower and its investments in Zambia and the theory that guided distributional differences between the two, we now see how inequality was baked into Zambia's energy infrastructure. Whereas balanced growth led to a deluge of benefits to Americans in the 1930s, a policy of unbalanced growth with hydropower have not resulted in a trickle down of material benefit to most Zambians.

Had the Washington Consensus never taken effect, manufacturing may have continued its upward trajectory and Zambia's economic growth may have been distributed more widely. With an improved environment of governance since the 2000s, manufacturing has been resurging and providing better employment and GDP returns per kWh of energy consumed than Zambia's largest consumer of energy, mining. However, as we saw in the previous chapter, energy supply to manufacturing reduced in 2015 and 2016, and with it, growth in manufacturing value-added. Not only did ZESCO provide manufacturing firms schedules for load-shedding, but firms also experienced unannounced power outages. The next chapter will review the literature on the impact of power outages on manufacturing as well as analyse data related to Zambia to discern clues on how the power outages of 2015 and 2016 may have affected manufacturing to result in the reduced growth that we have observed.

[30] For example Yugoslav firm Energoproject constructed the first stage of Kafue hydropower dam (Park, 1968, p. 16).

[31] The proof is in the export volume of Zambia's metal and ores, although the government does at least incentivise firms to process ores into blister copper (99.5% copper content) before exporting with high export tariffs on ores (Tembo, 2018, p. 42).

References

Ahmed, I. (2020a). Explaining Rwanda's prioritisation of rural electrification over rural clean drinking water through institutional path dependency. *Structural Change and Economic Dynamics, 54*(September), 186–201. https://doi.org/10.1016/j.strueco.2020.05.001

Ahmed, I. (2020b). The cost of power outages to Zambia's manufacturing firms, households and the climate. In *Grid reliability and utility operations.* The Applied Research Programme on Energy and Economic Growth (EEG).

Ahmed, I. (2021). The political economy of energy mix in hydropower dependent developing nations–a case study of Zambia. University College London. Available at: https://discovery.ucl.ac.uk/id/eprint/10121236/.

Ahmed, I., Baddeley, M., Coffman, D., Meikle, J., Oseni, M., & Sianjase, G. (2019). *The cost of power outages to Zambia's manufacturing sector.* https://www.theigc.org/wp-content/uploads/2019/08/Ahmed-et-al-2019-policy-brief-41408.pdf. https://doi.org/F-41408-ZMB-1

Ahmed, I., Parikh, P., Sianjase, G., & Coffman, D. (2020, December 4). The impact decades long dependence on hydropower in El Niño impact-prone Zambia is having on carbon emissions through diesel backup generation. *Environmental Research Letters, 15*(12) https://doi.org/10.1088/1748-9326/abb6a1

Alacevich, M. (2009). *The political economy of the World Bank—The early years.* Stanford University Press.

Bakker, K. (2013). Constructing 'public' water: the World Bank, urban water supply, and the biopolitics of development. *Environment and Planning D: Society and Space, 31*, 280–300. https://doi.org/10.1068/d5111

Barber, W. J. (1996). *Designs within disorder. Franklin D. Roosevelt, the Economists, and the Shaping of American Economic Policy, 1933–1945.* Cambridge University Press.

Barton, S. J. (2014). Sino-substitution: Chinese foreign direct investment in Zambia. *Journal of Chinese Economic and Foreign Trade Studies, 7*(2), 90–109. https://doi.org/10.1108/JCEFTS-08-2013-0025

Barton, S. J. (2016). *Policy signals and market responses—A 50 year history of Zambia's relationship with foreign capital.* (A. B. Bell, D. Coffman, & T. K. Moore, Eds.), 1st ed. Palgrave Macmillan.

Bayliss, K., & Pollen, G. (2019). *From Global to Local: Exploring the effects of the global energy policy paradigm in Zambia.* 119. Leeds. https://sri-workingpapers.leeds.ac.uk/wp-content/uploads/sites/67/2019/10/SRIPs-119.pdf

BBC. (2008). *Obituary: Zambia's Levy Mwanawasa, British Broadcasting Corporation News.* Retrieved November 23, 2018, from http://news.bbc.co.uk/1/hi/world/africa/7488623.stm

Beilfuss, R. (2012). *A risky climate for Southern African Hydro: Assessing hydrological risks and consequences for Zambezi River Basin Dams*. International Rivers. http://seine.oieau.fr/oieau.fr/sites/seine.oieau.fr.oieau.fr.eaudoc/files/32987.pdf

Bhorat, H. et al. (2017). *Understanding growth-income inequality interactions in Zambia*. https://www.theigc.org/wp-content/uploads/2018/04/Bhorat-et-al-2017-final-paper.pdf

Boos, A., & Holm-Müller, K. (2016). The Zambian Resource Curse and its influence on Genuine Savings as an indicator for 'weak' sustainable development. *Environment, Development and Sustainability, 18*(3), 881–919. https://doi.org/10.1007/s10668-015-9667-5

Cantore, N., et al. (2017). Manufacturing as an engine of growth: Which is the best fuel? *Structural Change and Economic Dynamics, 42*, 56–66. https://doi.org/10.1016/j.strueco.2017.04.004

Cardozo, A., et al. (2014). Growth, employment, diversification, and the political economy of private sector development in Zambia. In C. Adam, P. Collier, & M. Gondwe (Eds.), *Zambia: Building prosperity from resource wealth*. Oxford Scholarship Online. https://doi.org/10.1093/acprof:oso/9780199660605.003.0003

Castellano, A. et al. (2015). *Brighter Africa: The growth potential of the sub-Saharan electricity sector*. https://www.mckinsey.com/~/media/McKinsey/dotcom/client_service/EPNG/PDFs/Brighter_Africa-The_growth_potential_of_the_sub-Saharan_electricity_sector.ashx.

Coffman, D. (2013). Credibility, transparency, accountability, and the public credit under the Long Parliament and Commonwealth, 1643–1653. In D. Coffman, A. Leonard, & L. Neal (Eds.), *Questioning Credible Commitment—Perspectives on the rise of financial capitalism*. Cambridge University Press.

Coffman, D., Leonard, A., & Neal, L. (2013). *Questioning credible commitment—Perspectives on the rise of financial capitalism* (D. Coffman, A. Leonard, & L. Neal, Eds.). Cambridge University Press.

Coffman, D., & Neal, L. (2013). Introduction. In D. Coffman, A. Leonard, & L. Neal (Eds.), *Questioning credible commitment—Perspectives on the rise of financial capitalism*. Cambridge University Press.

Cooke, M. L. (1949, July 11). Down to Earth with Point Four. *New Republic*, 18–21.

Currie, L. (1950). *The basis of a development program for Colombia*. https://doi.org/10.2105/ajph.40.12.1567-a

Currie, L. B. (1979, August 31). *A talk by Lauchlin Currie: Hotel Tequendama*. Mimeo.

Díaz, J., Lüders, R., & Wagner, G. (2016). *Chile 1810–2010: La República En Cifras: Historical statistics*. First. Ediciones Universidad Católica of Chile. https://www.jstor.org/stable/j.ctt1h9cvqj

Dlamini, J., et al. (2016). Relationship between energy consumption and economic growth in South Africa: Evidence from the bootstrap rolling-window approach. *Energy Sources, Part B: Economics, Planning and Policy, 11*(7), 617–625. https://doi.org/10.1080/15567249.2013.843618

Douglas, W. O. (1951). Address before the General Assembly of the State of Tennessee. Appendix to the Congressional Record, 82nd Congress, 1 sess., p. 22 February.

Ekbladh, D. (2002). 'Mr. TVA': Grass-roots development, David Lilienthal, and the rise and fall of the Tennessee Valley authority as a symbol for U.S. overseas development, 1933–1973. *Diplomatic History, 26*(3), 335–374. https://doi.org/10.1111/1467-7709.00315

Energy Regulation Board of Zambia. (2015). *Energy Regulation 2014.*

Energy Regulation Board of Zambia. (2017). *Energy Sector Report 2016.* https://doi.org/10.13140/RG.2.1.3893.7043

Energy Regulation Board of Zambia. (2018). *Energy Sector Report 2017.* http://www.erb.org.zm/downloads/reports/esr2017.pdf

Finer, H. (1972). *The T.V.A.: Lessons for international application.* Da Capo Press.

Fukuyama, F. (2018). *Identity: The demand for dignity and the politics of resentment.* Macmillan.

Funder, M., et al. (2018). The politics of climate change adaptation in development: Authority, resource control and state intervention in rural Zambia. *The Journal of Development Studies. Routledge, 54*(1), 30–46. https://doi.org/10.1080/00220388.2016.1277021

Garner, R. L. (1952). *Letter to Rafael Delgado.* LBCP.

Garner, R. L. (1953). *Letter to Emilio Toro.* LBCP.

Gilpin, R. (2001). *Global political economy: Understanding the international economic order.* Princeton University Press.

Gondwe, M., & Pamu, E. M. (2014). Growth strategies for Zambia: Harnessing natural resource wealth for sustainable growth. In C. Adam, P. Collier, & M. Gondwe (Eds.), *Zambia: Building prosperity from resource wealth.* Oxford Scholarship Online. https://doi.org/10.1093/acprof:oso/9780199660605.003.0001

Haraguchi, N., Martorano, B., & Sanfilippo, M. (2019). What factors drive successful industrialization? Evidence and implications for developing countries. *Structural Change and Economic Dynamics, 49*, 266–276. https://doi.org/10.1016/j.strueco.2018.11.002

Hirschman, A. O. (1959). *The strategy of economic development.* Yale University Press.

Huxley, J. (1946). *TVA—Adventure in planning.* The Architectural Press.

IBRD. (1947). *Second Annual Report 1946–1947.*

IBRD. (1948). *Third Annual Report, 1947–1948.* https://doi.org/10.1007/978-1-349-26356-1_7

IBRD. (1950a). *Report and recommendations of the President to the Executive Directors on the proposed loan to Central Hidroelectrica de Caldas*. http://documents1.worldbank.org/curated/en/657441468022430739/pdf/multi0page.pdf
IBRD. (1950b). *Report and recommendations of the President to the Executive Directors on the proposed loan to Central Hidroelectrica del Rio Anchicaya*. Retrieved from http://documents1.worldbank.org/curated/en/619771468240592149/pdf/multi0page.pdf
IBRD. (1950c). *Technical report of the Paulo Afonso Hydroelectric Project in Brazil*. https://projects.worldbank.org/en/projects-operations/projects-list?lang=en&searchTerm=&countrycode_exact=BR
IBRD. (1951a). *Report and recommendations of the President to the Executive Directors on the proposed loan to Central Hidroelectrica del Rio Lebrija*. http://documents1.worldbank.org/curated/en/216361468026996331/pdf/multi0page.pdf
IBRD. (1951b). *Report on Cuba—Findings and recommendations of an economic and technical mission*. https://doi.org/10.2307/2509624
IBRD. (1951c). *Sixth Annual Report 1950–1951*.
IBRD. (1952a). *Technical report on the Southern Rhodesian power expansion project*.
IBRD. (1952b). *The Seyhan River multipurpose project in Turkey*. http://documents1.worldbank.org/curated/en/855141468120269687/pdf/multi-page.pdf
IBRD. (1953a). *Report and recommendations of the President to the Executive Directors on the proposed loan to the Companhia de Electricidade do Alto Rio Grande (CEARG) and to the Centrais Eletricas de Minas Gerais S.A. (cemig) in the United States of Brazil (Itutinga Hy*. https://projects.worldbank.org/en/projects-operations/projects-list?lang=en&searchTerm=&countrycode_exact=BR
IBRD. (1953b). *The economic development of Nicaragua*.
IBRD. (1961a). *Appraisal of The Volta River Hydroelectric Project Ghana*. http://documents1.worldbank.org/curated/en/684761468030834258/pdf/multi-page.pdf
IBRD. (1961b). *Report and recommendations of the President to the Executive Directors on a proposed loan to the Ethiopian electric light and power authority*. https://doi.org/10.1017/S0020818300024826
IBRD. (2020). *Subscriptions and voting power of member countries*. http://pubdocs.worldbank.org/en/795101541106471736/IBRDCountryVotingTable.pdf
IBRD and IDA. (1970). *Appraisal of Kariba North Hydroelectric Project Zambia*. http://documents.worldbank.org/curated/en/917071468334477402/pdf/multi-page.pdf

IBRD and IDA. (1971). *Appraisal of the Kamburu Hydroelectric Project Tana River Development Company Limited Republic of Kenya*. https://doi.org/10.1017/S0020818300024826

IBRD and IDA. (1973). *Appraisal of Kafue Hydroelectric Project (Stage II) Zambia*. http://documents.worldbank.org/curated/en/824421468167973026/pdf/multi-page.pdf

Jerven, M. (2013). *Poor numbers how we are misled by African development statistics and what to do about it*. Cornell University Press (Cornell studies in political economy).

Jiang, L., et al. (2020). Rural household energy consumption of farmers and herders in the Qinghai-Tibet Plateau. *Energy, 192*, 116649. https://doi.org/10.1016/j.energy.2019.116649

Johnson, L. B. (1965). *Peace without conquest*. C-SPAN. https://www.c-span.org/video/?153274-1/reel-america-president-lyndon-johnson-vietnam-war-1965

Kaldor, N. (1970). The case for regional policies. *Scottish Journal of Political Economy, 17*(3), 337–348. https://doi.org/10.1111/j.1467-9485.1970.tb00712.x

Kaldor, N. (1984) *Causes of growth and stagnation in the world economy*. Cambridge University Press (Raffaele Mattioli Series). https://doi.org/10.1017/CBO9780511559709

Kapur, D., Lewis, J. P., & Webb, R. C. (1997). *The World Bank: Its first half century*. Brookings Institution Press. http://documents.worldbank.org/curated/en/313081468322727631/History

Kuznets, S. (1955). Economic growth and income inequality. *The American Economic Review, 45*(1), 1–28. https://www.jstor.org/stable/1811581

Lusaka South Multi Facility Economic Zone. (2018). *Investors in the Zone, Lusaka South Multi Facility Economic Zone website*. Retrieved November 26, 2018, from http://www.lsmfez.co.zm/

MacroTrends. (2020). *Copper Prices—45 Year Historical Chart, MacroTrends.net*. Retrieved June 6, 2020, from https://www.macrotrends.net/1476/copper-prices-historical-chart-data

Marconi, N., de Reis, C. F. B., & de Araújo, E. C. (2016). Manufacturing and economic development: The actuality of Kaldor's first and second laws. *Structural Change and Economic Dynamics, 37*, 75–89. https://doi.org/10.1016/j.strueco.2015.12.002

Mason, E. S., & Asher, R. E. (1973). *The World Bank since Bretton Woods*. The Brookings Institution.

McKinney, J., et al. (2020). *Sustainability of China's overseas special economic zones: Zambia, National Geospatial-Intelligence Agency*. Retrieved August 9, 2020, from https://www.tearline.mil/public_page/sustainability-of-chinas-overseas-special-economic-zones-zambia/

McMillan, M. S., & Rodrik, D. (2011). Globalization, structural change and productivity growth. https://doi.org/10.3386/w17143

McNamara, R. S. (1966, May 18). Security in the contemporary world. Address of the US Secretary of Defense to the American Society of Newspaper Editors.

Mfula, C. (2010a). First Quantum's Zambia mine hit by power outage. *Reuters.* https://www.reuters.com/article/update-2-zambia-mining-power-idAFLDE6180X820100209

Mfula, C. (2010b, August 21). Singapore firm to spend $108 mln on Zambia coal mine. *Reuters.* https://uk.reuters.com/article/mining-zambia/singapore-firm-to-spend-108-mln-on-zambia-coal-mine-idUKLDE67K09320100821

Morrell, D. (1987). *Indictment: Power and politics in the construction industry.* Faber and Faber Ltd..

Moyo, D. (2013). *Winner take all: China's race for resources and what it means for us.* Penguin.

Mulanda, S. M., & Punt, C. (2020). *Structural change and economic dynamics structural characteristics of Zambia's agricultural sector and the role for agricultural policy : Insights from SAM based modelling.* Stellenbosch.

Ncube, M., Lufumpa, C. L., & Kararach, G. (2017). Introduction: Infrastructure in African development. In M. Ncube & C. L. Lufumpa (Eds.), *Infrastructure in Africa—Lessons for future development.* Policy Press.

North, D. C., & Weingast, B. R. (1989). Constitutions and commitment: The evolution of institutions governing public choice in seventeenth-century England. *The Journal of Economic History, 49*(4), 803–832. https://doi.org/10.1017/S0022050700009451

OECD. (2019). Aid (ODA) by sector and donor, OECD website.

Olivier, R. (1961). Transcript of interview with Burke Knapp.

Park, S. F. (1968). *Basic data on the economy of Zambia.* https://play.google.com/books/reader?id=K_ZhUrU5lfoC&hl=en_GB&pg=GBS.PA1

Piketty, T. (2014). *Capital in the twenty-first century* (A. Goldhammer, Ed.). The Belknap Press of Harvard University Press.

Reinert, E. S. (2019). German economics as development economics: From the thirty years' war to World War II. In R. Kattel (Ed.), *The visionary realism of German economics: from the thirty years' war to the cold war* (pp. 15–36). Anthem Press. https://doi.org/10.2307/j.ctvcb5979

Resnick, D., & Thurlow, J. (2014). *The political economy of Zambia's recovery: Structural change without transformation?*http://ebrary.ifpri.org/utils/getfile/collection/p15738coll2/id/128012/filename/128223.pdf

Reuters. (2013, January 7). Zambia's Sata says high maize meal price could spark riots. *Reuters.* https://www.reuters.com/article/zambia-riots-idUSL5E9C75HI20130107

Rodrik, D. (2016). Premature deindustrialization. *Journal of Economic Growth, 21*(1), 1–33. https://doi.org/10.1007/s10887-015-9122-3

Roosevelt, F. D. (1945, February). Message to congress on the Bretton woods agreements. The American Presidency Project, 12. https://www.presidency.ucsb.edu/node/210026

Rosenstein-Rodan, P. N. (1943). Problems of industrialisation of Eastern and South-Eastern Europe. *The Economic Journal, 53*(210/211), 202–211. http://www.jstor.org/stable/2226317

Rostow, W. W. (1960). *The stages of economic growth: A non-communist manifesto* (3rd ed.). Cambridge University Press.

Schlesinger, A. J. (1949). *The Vital Center: The politics of freedom*. Houghton Mifflin.

Scott, J. C. (1998). *Seeing like a state—How certain schemes to improve the human condition have failed*. Yale University Press.

Sharma. (2010). *Note to users*. https://doi.org/10.1177/001088048102200214

Sharma, P. A. (2017). *Robert McNamara's other war—The World Bank and International Development*. University of Pennsylvania Press.

Simpasa, A. et al. (2013). *Capturing mineral revenues in Zambia: Past trends and future prospects capturing mineral revenues in Zambia: Past trends and future prospects*. https://www.un.org/en/land-natural-resources-conflict/pdfs/capturing-mineral-revenues-zambia.pdf

Solow, R. M. (1956). A contribution to the theory of economic growth. *The Quarterly Journal of Economics, 70*(1), 65–94. http://www.jstor.org/stable/1884513

Staples, A. L. S. (2006). *The birth of development: How the World Bank, Food and Agriculture Organisation and World Health Organisation changed the world, 1945–1965*. The Kent State University Press.

Steel News. (2011). Konkola Copper Mines to invest USD 33 million for diesel generators. *Steel Guru*. https://steelguru.com/steel/konkola-copper-mines-to-invest-usd-33-million-for-diesel-generators/242497

Stiglitz, J. E. (2017). Promoting Sustainable Industrial Policies. In *Industrialize Africa: Strategies, Policies, Institutions and Financing*, pp. 11–15 African Development Bank Group. https://www.afdb.org/en/news-and-events/industrialize-africa-strategies-policies-institutions-and-financing-17570/

Syndigate. (2011, June 17). KCM deploys 3 diesel generators at Konkola Deep Mining Project. *MENA Report*. https://search-proquest-com.libproxy.ucl.ac.uk/docview/872316305?rfr_id=info%3Axri%2Fsid%3Aprimo

Szirmai, A. (2012). Industrialisation as an engine of growth in developing countries, 1950–2005. *Structural Change and Economic Dynamics, 23*(4), 406–420. https://doi.org/10.1016/j.strueco.2011.01.005

Szirmai, A., & Verspagen, B. (2015). Manufacturing and economic growth in developing countries, 1950–2005. *Structural Change and Economic Dynamics, 34*, 46–59. https://doi.org/10.1016/j.strueco.2015.06.002

Tembo, B. (2018). *Strategic investment decisions in Zambia's mining sector under a constrained energy system*. University College London. https://discovery.ucl.ac.uk/id/eprint/10048151/

Thurlow, J., & Wobst, P. (2007). The role of agriculture in pro-poor growth: Lessons from Zambia. In M. Grimm, S. Klasen, & A. McKay (Eds.), *Determinants of pro-poor growth—Analytical issues and findings from country cases* (pp. 219–241). Palgrave Macmillan.

Tobey, R. C. (1996). *Technology as freedom: The New Deal and the electrical modernization of the American home*. University of California Press.

Transparency International. (2020). *Corruption Perceptions Index 2019, Corruption Perceptions Index 2019*. Retrieved March 2, 2020, from https://www.transparency.org/country/ZMB

U.S. Census Bureau. (1999). Table No. 1430. Employment Status of the Civilian Population: 1929 to 1998. In *Statistical abstract of the United States*. U.S. Census Bureau. https://www.census.gov/prod/99pubs/99statab/sec31.pdf

UN. (2008). International standard industrial classification of all economic activities. New York. https://unstats.un.org/unsd/publication/seriesm/seriesm_4rev4e.pdf

United Nations. (1962). *International Trade Statistics 1900–1960, UN Trade Statistics*. https://unstats.un.org/unsd/trade/data/tables.asp

Whitworth, A. (2014). Energy policy. In C. Adam, P. Collier, & M. Gondwe (Eds.), *Zambia: Building prosperity from resource wealth*. Oxford Scholarship Online. https://doi.org/10.1093/acprof:oso/9780199660605.003.0011

Wobst, P., & Thurlow, J. (2005). The road to pro-poor growth in Zambia: Past lessons and future challenges. In *Proceedings of the German Development Economics Conference, Kiel 2005 / Verein für Socialpolitik, Research Committee Development Economics, No. 37*. International Food Policy Research Institute. http://hdl.handle.net/10419/19828

World Bank. (2016). World Bank's First Development Loans to Chile, 1948, (057), pp. 1–5. http://documents.worldbank.org/curated/en/308691468185347709/pdf/104690-WP-PUBLIC-2007-01-World-Banks-First-Development-Loans-to-Chile.pdf

World Bank. (2017b). *Investor brief*. DC. http://pubdocs.worldbank.org/en/284471507050951684/investor-brief-worldbank-2017.pdf

World Bank. (2017c). *World development indicators—Zambia*. Data Bank. https://data.worldbank.org/country/zambia

World Bank. (2018). World Bank projects and operations in Zambia. World Bank. https://doi.org/10.1007/s10461-016-1645-x

World Bank. (2019a). *World development indicators: Zambia*. World Bank. https://data.worldbank.org/country/zambia

World Bank. (2019b). *World development indicators—Chile*. https://data.worldbank.org/country/chile

World Bank. (2020a). *Metadata glossary, databank*. Retrieved October 24, 2020, from https://databank.worldbank.org/metadataglossary/wdi-database-archives-(beta)/series/NV.IND.TOTL.CD#:~:text=Longdefinition,electricity%2Cwater%2Candgas

World Bank. (2020b). *Projects in Colombia, World Bank*. Retrieved October 17, 2020, from https://projects.worldbank.org/en/projects-operations/projects-list?lang=en&searchTerm=&countrycode_exact=CO

World Bank. (2020c). *Projects in Ethiopia, World Bank*. Retrieved October 17, 2020, from https://projects.worldbank.org/en/projects-operations/projects-list?lang=en&searchTerm=&countrycode_exact=ET

World Bank. (2020d). *Projects in Turkey, World Bank projects & operations*. Retrieved October 18, 2020, from https://projects.worldbank.org/en/projects-operations/projects-list?lang=en&searchTerm=&countrycode_exact=TR

World Bank. (2020e). *Zambia Trade Summary 2018 Data, World Integrated Trade Solution*. Retrieved April 11, 2020, from https://wits.worldbank.org/CountryProfile/en/Country/ZMB/Year/LTST/Summary

World Bank. (2020f). *Zambia world development indicators*.

Zambia Sugar. (2017). *2016 Annual Report*. https://africanfinancials.com/document/zm-zmsg-2016-ar-00/

Zeng, D. Z. (2016). *Multi-facility economic zones in Zambia: Progress, Challenges and possible interventions*. http://documents1.worldbank.org/curated/en/720981495115586647/pdf/115143-WP-PUBLIC-Feb-2016-GTCCS-ZambiaMFEZ.pdf

Zesco Ltd. (2018). 2017 Integrated report.

ZICTA and CSO. (2018). *2018 National survey on access and usage of information and communication technologies by households and individuals—A demand side assessment of access and usage of ICTs in Zambia*. https://www.zicta.zm/Downloads/publications/2018ICTSurvey_PreliminaryReport.pdf

CHAPTER 3

The Impact of Power Outages in the Literature

The previous chapter shows how Zambia's system of energy provision prioritised mining, beneficiation and manufacturing to kick-start the motor of Zambia's economy. This chapter examines the literature on the impact of power outages on manufacturing, and on Zambian sectors of the economy and manufacturing in particular. It also uses reports and statistics to supplement this understanding and identifies gaps in the literature that primary data collection can plug.

Studies assessing the value of unsupplied electricity can come both on the basis of aggregate macro data as well as on the basis of individual people or firms, forming micro analysis (Bental & Ravid, 1982).

3.1 Macro Versus Micro Analysis on the Impact of Power Outages Across Sectors

The limitations of macro-level analysis have long been established. Macro analysis looks at the long-run effects outages have on long-run growth through the lens of the neoclassical Solow growth model (Solow, 1956; Andersen & Dalgaard, 2013). To illustrate how macro analysis failed, Bental and Ravid (1982) pointed to a Chilean study (Jaramillo & Skoknic, 1973), where the authors inferred that industries which were least electricity intense would have the largest loss per unit of reduced supply. Another pitfall with macro studies according to Bental and Ravid is that they

I. Ahmed, *The Political Economy of Hydropower Dependant Nations*,
https://doi.org/10.1007/978-3-030-71266-2_3

estimate the average cost of unsupplied electricity, whereas the relevant estimate would be marginal cost. They also often fail to take into substitution away from electricity-intensive activities (Bental & Ravid, 1982).

Andersen and Dalgaard returned to a macro analysis approach in 2013. Using outages as a binary variable in a logarithmic regression model, Andersen and Dalgaard estimated that a 1% increase in the average frequency of outages in a month reduces long-run GDP by almost 3% for a sample of 39 Sub-Saharan countries, including Zambia (Andersen & Dalgaard, 2013). If all African countries had experienced South Africa's power quality, they state, their GDP per capita growth rate would have been increased by 2 percentage points. The problems with this macro analysis are several. Recognising that GDP is likely to be plagued by non-random measurement error in Africa, they produced 'adjusted' real GDP per capita growth rates by looking at the growth in nightlights using satellite data. But non-growth of luminance is not necessarily representative of non-growth of GDP: it may be that unequal growth results in those who already had nightlights are getting richer, but have no more lights to shine, while those without nightlights previously are not getting richer and still do not have nightlights.

The weakness with micro studies is often that they rely on questionnaires prone to inaccurate responses. Respondents may have reason to overstate the damage done of power outages (Bental & Ravid, 1982). Beenstock et al. (1997) observed a number of instances respondents to surveys reported losses that the authors considered unreasonably large. They interpreted these overstatements as 'protest' responses; the interviewees expressing their discontent with the power utility. Another pitfall of subjectively answered surveys is that they could be prone to loss aversion.[1] This may explain why a number of studies such as Pasha et al.'s in Pakistan (1989) and Beenstock et al.'s in Israel (1997) concluded that firms overstated their losses. Similarly, perceived losses do not necessarily match actual losses if framed in a particular way, contrary to rational expectation.

In reviewing the literature on whether individuals' willingness to abandon electricity is greater than their willingness to pay for it, Plott and

[1] Behavioural economist Richard Thaler (1980) observed that people suffer from loss aversion: the loss of utility from giving up a valued good is greater than the gain in utility from acquiring the good.

Zeiler (2005) point out that there is not a consensus.[2] They themselves used procedures which resulted in no difference between willingness to accept and willingness to pay, calling into question the applicability of prospect theory.[3] The key, they wrote, is to avoid misconceptions in willingness to pay questions and provide some sort of incentive for truthful revelation of valuations. Willingness to pay an additional premium for reliable electricity could still therefore be potentially useful for gauging the marginal cost of backup power. The loss caused by outages is converted into a positive extra price that they would have to pay. Where respondents are firms and not individuals, the 'house money effect' should not come into play since respondents themselves would not be paying the higher tariffs that their firms would. Additionally, while willingness to pay could be used to gauge the marginal cost of backup generation, to avoid the problem of subjectivity associated with micro studies, observations can be used to ascertain the marginal cost of power outages.

Steinbuks and Foster (2010) took a micro analysis approach to investigating the impact of power outages across 25 African countries, Zambia among them. Using World Bank enterprise survey data from 2002 to 2006, the surveys drew on information collected across the manufacturing and services sectors. One of their identification methods used results of reported loss of sales due to power outages. This is open to both the loss aversion and protest critiques, and is not resolved by seeking how much more respondents would have been willing to pay for more reliable energy.

One noteworthy finding of theirs is their calculation of Zambia's weighted average cost of electricity for firms using backup generators. Adding the variable cost of backup energy to the average capital cost of backup energy, and weighting that by 100% of the time of power outages in relation to non-outage hours, which they ascertained from the World

[2] Thaler's loss aversion theory, also known as 'endowment effect theory', would predict that individuals would state a minimum amount for which they would be willing to abandon electricity as greater than the maximum amount they would be willing to pay for it (Plott & Zeiler, 2005, pp. 530–531). This runs contrary to orthodox economic 'expected utility theory' which predicts that under conditions of sufficiently smooth preferences and the absence of wealth effects, the magnitudes of willingness to buy and willingness to forgo the electricity should be the same (Plott & Zeiler, 2005, pp. 530–531; Baddeley, 2013, p. 15).

[3] Thaler's loss aversion theory builds upon Kahneman and Tversky's prospect theory which explains why utility theory does not explain real-world choices made by people which are inconsistent and intransitive (Baddeley, 2013, p. 139). In prospect theory, value is assigned to gains and losses rather than to final outcomes, and probabilities are replaced by decision weights (Kahneman & Tversky, 1979).

Bank enterprise survey data, they calculated Zambia's weighted average cost of electricity to be USD 0.02/kWh greater than the cost of ZESCO energy (Steinbuks & Foster, 2010). Their method has one apparent weakness and rests on an assumption that requires further examination. The weakness is that it is incorrect to divide the capital cost of backup generation by the hours of use each year, since generators last longer than a year. If anything, they should have divided the capital cost by the expected lifespan of the generator in terms of hours. The assumption that requires further examination is that firms with backup generators use backup generators 100% of the time and to capacity during power outages, an assumption shared by earlier literature such as that by Bental and Ravid (1982). The problem with this is that this may not even have been rational behaviour for the profit-maximising firm: the variable gross profit of production may have become negative with the use of backup generation. Firms may also have lacked the working capital necessary to run generators due to higher running costs, especially fuel costs. It may also not have been a policy followed because the firms' management suffered from heuristic behavioural biases, miscalculated the costs involved or misjudged the length of anticipated outages *ex ante*.

Across Africa, the Word Bank's latest enterprise survey data reveals that 41% of surveyed Sub-Saharan African firms identified electricity as the major constraint to their businesses, compared with 26% identifying transportation as a major constraint. Seventy-five percent of firms experienced electrical outages, which on average experienced 9 outages in a typical month, and each outage lasted on average 5.8 hours. Outages on average cost 8.5% of annual sales (a subjective estimate subject to loss aversion and protesting). Fifty-three percent of firms owned or shared generators. When generators were used, the average proportion of energy generated was 28% (World Bank, 2018).

3.2 The Impact of Power Outages on Large Manufacturing Firms

The literature on the impact of power outages to the manufacturing sector in general is rich—studies have been carried out since at least 1948, in Sweden (Bental & Ravid, 1982)—and there are a number of studies that hone in on the impact of outages on Zambia's manufacturing sector in particular. None of these studies on Zambia, however, study the impact of the severe outages of 2015 and 2016 on large manufacturing firms, including those published after 2015.

3.2.1 Reduced Productivity

Using World Bank Enterprise Survey data collected from 2001 to 2005 for 1000 firms in ten Sub-Saharan countries, including Zambia, Arnold et al. (2008) found that firms in regions with more frequent power outages are less productive than others. While the regression result which pitted days with power outages against total factor productivity that gave rise to this conclusion was only significant at the 10% level using ordinary least squares, and the R-squared for the ordinary least squares model was only 30%, this result was significant at the 1% level using the semi-parametrically estimated firm productivities per Olley and Pakes (1996) with a 48% R-squared for the model. Moreover, the coefficient on the share of firms that own a generator to be significant and positive at the 5% level in both models.

Similar in aim to Arnold et al.'s study, Escribano et al. (2010) used World Bank investment climate surveys of 26 African countries carried out in 2002–2006 to consider the impact of poor quality of power, as measured by number and average duration of power outages and the use of a generator on total factor productivity. They calculated the contribution of the average duration of power outages to average log total factor productivity as 9%, though the level of significance of this result was not shared (their publication was a World Bank working paper and not a peer-reviewed journal).

Pre 2015/2016 power outages in Zambia accounted for 5.5% of sales value for the World Bank's surveyed firms. The loss was felt mostly greatly by medium-sized firms employing 20–99 people, which lost 6.9% of value and felt least by large firms employing more than 100 people (accounting for only 4.7% of sales). This was the case even though they reportedly faced 8.7 power outages in a typical month, compared to the 5.4 outages per month by medium-sized businesses (World Bank and International Finance Corporation, 2014, p. 14). The reasons for this were not given.

Also using World Bank data (enterprise survey data of 720 firms collected between 2012 and 2014) for Zambia, Sichone et al. (2016) found that load shedding had a statistically significant (at the 5% level) and negative impact on manufacturing output in Lusaka.

Generally, the mechanism by which productivity reduces with power outages is that electronically operated capital ceases to be operational without power. Worker productivity falls. Outages result in restart costs, lost output and sales for industries dependent on electricity, damaged equipment, destruction of raw materials, loss in quality of production and lost reputation—particularly for export-oriented firms, such as reduced rankings on export markets' reliability criteria (Beenstock et al., 1997; Steinbuks & Foster, 2010). Even when making up for lost load, firms lose with higher wear and tear as they use their machinery more intensely and with higher staff costs associated with over-time shifts. There are thus welfare losses (Kessides, 1993) and economic growth dampens (Eberhard et al., 2008; IMF, 2008, chap. IV). However, Allcott et al. (2015) noted that answering how electricity outages affect productivity in the manufacturing sector can be difficult to answer first of all because the quality of data collection in countries in which outages occur is not high, but also because of endogeneity: rapid economic growth itself may be causing loss in productivity because of the increase in electricity demand that leads to shortages, and poor institutions could lead to insufficient power supply and reduce productivity.

3.2.2 Subsector Impact and Multiplier Effects

Pasha et al. (1989) first considered the idea that the impact of power outages on an industry could have a downstream effect. Using values calculated by others (Naqvi & Ahmed, 1986) for the coefficients of variables in a truncated[4] formula for aggregate demand, they calculated a static multiplier of 1.34 for all industries. This implied that industrial outage costs could increase as much as 34% when one considered the downstream impact, regardless of manufacturing subsector. Given different manufacturing subsectors' different outputs, a static multiplier value is unlikely to be a good measure for any given subsector.

Chen and Vella (1994) went further by disaggregating multipliers by manufacturing subsectors, to assess the impact of power outages on a particular subsector on the rest of the economy (Chen & Vella, 1994). Their research provided the calculations to be performed and stating that a FORTRAN programme was written for computing the input-output framework in its application for the Taiwanese economy. It was unclear

[4] Government expenditure and net exports were omitted.

how they ascertained the required data, how rich it was, and what problems they faced with it. The value of the analysis for informing policy is limited by the fact that it does not consider the volume of value-added by industries. It is useful to know that for every Taiwanese dollar the steel and iron industry supplied its products, downstream value-added was NT$0.917, but it would also have been good to know what total value the steel and iron industry added to the Taiwanese economy, not to mention the total employment it brought. At least Pasha et al. considered the variation in outage cost per kilowatt hour by industry, and the knock-on effects for Pakistan's exports. Continuous-process industries and those more vulnerable to spoilage experience greater losses, they found. To minimise loss, therefore, power-feeders should prioritise these industries (Pasha et al., 1989). Chen and Vella, on the other hand, saw their methodology as a possibly useful one for figuring out the electricity customers to cut first (Chen & Vella, 1994, pp. 1064–1065), which, given its incompleteness, makes their suggestion a dangerous one.

In China, Fisher-Vanden et al. (2015) found that across all industries, material input expenditures increased by 13% in response to electricity shortages since 1999. They found the largest effects in the wood products, chemicals, food, metal and textiles industries. However, this was offset by a 5% reduction in unit cost due to savings in other inputs and small total factor productivity improvements.

For Zambia, using World Bank enterprise survey data with 720 observations, Sichone et al. (2016) found that the impact of power outages was only significant and negative for food and metal fabrication, associated with a 1.9% and 1.4% reduction in sales respectively. The impact was not statistically significant for furniture and wood processing, garments and leather and non-metallic minerals. They posit that the insignificant impact of self-generation on wood processing, garments and leather and non-metallic minerals subsectors is because they are less dependent on electricity for production (Sichone et al., 2016).

For the food subsector and metal fabrication, when production starts, Sichone said that there are some processes that should not be interrupted (Ahmed, 2021). Because spoilage costs exceed the cost of running self-generation, to hedge against spoilage costs, firms in these sectors will run their generators even when the grid is providing power (Ahmed, 2021). In one instance, he saw a glass factory's clinker solidify in pipes, resulting in the need to replace pipes. He also told us that ZESCO would not always keep to its load-shedding schedules.

Running along the theme of discerning the impact of power outages on different customers, Allcott et al. (2015) suggested that offering interruptible electricity contracts at a discount could substantially reduce the impact of shortages. This bottom-up approach, rather like the purchase of generators, would allow for customers to determine for themselves their value of reliable electricity. Unlike Chen and Vella's suggestion, which aspires for a central decision-making body to figure out how to maximise social economic benefit (Chen & Vella, 1994, p. 1067), it relies on the 'invisible hand' to figure out how to reduce the costs of shortages—if not achieve the socially optimal result.

3.2.3 Peak Versus Off-Peak Tariffs

Pasha et al. found that 28% of firms surveyed not operating 24 hours per day would be willing to shift their work timings if lower off-peak tariffs were offered, ranging from 11% in the metal industry to 52% in the paper and wood industry (Pasha et al., 1989). Pasha et al. also suggested seasonal variation in the level of tariffs generally, because hydropower-dependent Pakistan in 1989, like Zambia today, saw a difference in the incidence of outages between winter and the early summer months (Pasha et al., 1989).

3.2.4 Duration of Outages

Fewer long-duration interruptions are less damaging to industry than several interruptions of very short duration (Pasha et al., 1989; Diboma & Tamo Tatietse, 2013). This is because short outages are generally not anticipated and second, because the spoilage of inventory and machinery and restart-up costs are linked to frequency and not duration. Spoilage costs spread over a longer duration result in less spoilage per kilowatt hour lost.

Willis and Garrod (1997) defined short interruptions as those in which the voltage falls below 0.1 relative units for a period shorter than or equal to 1 minute. Long interruptions are those in which the voltage falls to zero for a period above a minute. Using collected survey data, Beenstock et al. (1997) found a U-shaped relationship between the loss per unsupplied unit of energy and the duration of the outage. Restart and equipment damage costs per kilowatt hour unsupplied vary inversely with duration, whereas output losses increase with duration. Their data suggests that the cost per kilowatt hour unsupplied reaches a minimum at 45 minutes for the firms that they surveyed in the mid-1990s in Israel. The U-shaped relationship between cost per unit of energy and duration perhaps explains Diboma's observation (2013) that businesses prefer longer, infrequent outages to shorter frequent outages.

To be able to adjust to outages (in terms of shift timings and work days), Pasha et al. (1989) found that surveyed firms in Pakistan said they needed 6–13 hours of lead time between notification and outage. Surveyed firms in Cameroon, meanwhile, estimated that their costs could be reduced by a fifth to a third if they were given reliable advance interruption notices (Diboma & Tamo Tatietse, 2013).

3.2.5 *Coping Mechanisms Used by Manufacturing Firms to Mitigate Against Power Outages*

The literature identifies myriad ways in which firms can respond to unreliable electric supply:

1. Insuring against loss of power (or not, as is the case in 25 African countries including Zambia according to Steinbuks & Foster, 2010).
2. Substituting away from electricity to other factors of production in the long-term (Solow, 1956; Bental & Ravid, 1982; Baddeley, 2003).
3. Negotiating flexible labour contracts to counteract reductions in labour productivity (as was the case in Pakistan: Pasha et al., 1989).
4. Taking measures to reduce equipment damage, such as installing voltage regulators, protection switch gears, managing loads, conserving energy (Sanghvi, 1982).
5. Carrying larger process inventories (Sanghvi, 1982).
6. Substituting techniques (Samuelson, 1962; Baddeley, 2003).
7. Changing business activity and allowing firms in other locations with more reliable power supply to take over the production activity, or alternatively changing location to an area of more reliable energy supply. Fisher-Vanden et al. (2015) found that firms in Chinese regions where electric power became scarcer shifted from 'makers' to 'buyers' of intermediate goods for production.

8. Privately generating their own power. Studies show that firms that self-generate electricity suffer smaller outage losses (Arnold et al., 2008; Steinbuks & Foster, 2010). Self-generation accounts for more than a fifth of generation capacity in some African countries (Foster & Steinbuks, 2009). However, while self-generation reduces outage losses, Oseni and Pollitt (2015) found that self-generation might not necessarily make a firm suffer smaller unmitigated outage losses. In calculating the unmitigated loss, Oseni and Pollitt calculated it as equal to the loss amount less the cost of self-generation. They should have also added to this the cost of self-generating energy above the cost of purchasing energy from the grid.

   Variables correlated with generator installation included size, number of days without power, whether they were foreign and whether they were export-facing, Reinikka and Svensson (2002) in Uganda. Similarly, using business survey data from 25 African countries, Steinbuks and Foster (2010) found that the size of a firm and whether it exports play more important roles than reliable supply of energy in the decision to invest in a backup generator. Allcott et al. (2015) explain the reason for size of firm being a correlate with owning a generator in terms of the rational firm: there are substantial economies of scale in generator costs.

   Because generators operate a small fraction of the time energy is used, they do not greatly affect the average cost to industry. Indivisibilities occur in backup investment (Beenstock et al., 1997). Because of this, it may cost a firm the same to invest in backup generation for 110% of its energy needs as it costs to back up 97% of its generation needs, and so it may decide not to back up against 100% of losses.
9. Reducing output (Adenikinju, 2003).
10. Changing business and possibly industry. This would be difficult to observe, and we have not seen studies tackling this issue. We would need to take into account survival bias as we interviewed manufacturing firms in Zambia following the major outages of 2015 and 2016.

3.2.6 *Use of Backup Generators to Calculate Observed Floors for Marginal Costs of Power Outages*

To avoid the problem of subjectivity associated with micro studies discussed above, observations can be used to ascertain the marginal cost of power outages.

Bental and Ravid (1982) and others subsequently, such as Adenikinju (2003) and Steinbuks and Foster (Steinbuks & Foster, 2010) argued that because acquiring backup generation is not costless (Adenikinju, 2003), the formulation of marginal cost of outages should include in addition to the variable cost the capital expenditure of purchasing a given capacity, divided by the hours of use in a year. This is an incorrect formulation. Marginal cost is the cost of producing one additional unit of a product or service. The acquisition of backup generation is only relevant at the time of acquisition, and is thereafter a sunk cost. For this reason, the time of acquisition of generators would be interesting to know for Zambia, as it would inform when power outages were bad enough to justify the capital expense of a backup generator, an apparent gap in the literature.

Farquharson et al. (2018) took the correct approach to calculating the marginal cost of backup generation, although they called it 'premium cost', and they likely calculated the wrong value as a result of using the wrong input value. They calculated that consumers in Zambia pay a cost premium for backup generation of USD 0.53/kWh. They calculated this as the difference between diesel backup generation and the price of grid electricity. This is the correct formulation of marginal cost—the additive cost, necessarily considering the cost of electricity that would have otherwise been paid. However, Farquharson et al. base their calculation on a 2014 cost of USD 0.59/kWh for diesel generation exclusive of capital expenditure, operating and maintenance expenses, which is far higher than what would have been expected. Steinbuks and Foster found that the average variable cost of self-generated electricity in Zambia was USD 0.27/kWh in 2010 (Steinbuks & Foster, 2010), personal experience negotiating power purchase agreements in Rwanda suggested that the cost was circa USD 0.29–0.34/kWh (personal experience in Rwanda, 2012–2016).

3.2.7 Predictors of Possession of Generators

Sichone's research (remarked on above) suggests that certain subsectors are more impacted than others because of their greater dependence on energy and vulnerability to power outages. Based on this, the subsector of manufacturing that a firm falls into could be a predictor of whether it owns a generator.

Steinbuks and Foster (2010) investigated what drives firms to generate their own power using data from the World Bank's Enterprise Survey Database comprising 8483 operating firms sampled between 2002 and 2006 across 25 African countries, including Zambia. Using the binary choice model of Reinikka and Svensson (2002), they looked at the probability of firms investing in generators by firm characteristics, using the probit method. They found that the probability of finding a generator on the premises increases by nearly 50% as one moves from firms employing less than ten employees to firms employing over 500 employees; and that the probability of having a generator remains high even where power supply is reliable. They found statistically significant relationships for several firm characteristics: days of power outages, size in terms of employees, age, export orientation and sector. The following is a rank of magnitude of coefficient for statistically significant sector variables for most likely to own a generator: hotels and restaurants, food and beverages, chemicals and pharmaceutics, non-metallic and plastic materials, metals and machinery, wood and furniture—mining did not have a statistically significant coefficient.

Oseni and Pollitt (2015) also attempted to identify predictors of firms' decision to invest in self-generation. Unlike Steinbuks and Foster, they did not find either days of power outages or age to be statistically significant predictors. They did find size in terms of employment and various sectors to be statistically significant. In terms of statistically significant sectors, they ranked as follows in terms of magnitude of coefficient: other services, food and beverages, electronics, chemical and pharmaceuticals, non-metals, retail, garment and textiles. Besides adding electricity consumption to their model that Steinbuks and Fuchs did not (and hence had a lower R-squared), they also added countries, further contributing to their R-squared. With their data coming from a dataset that was published in 2012, they found that Zambia had a statistically significant result at the 5% level, but with a negative coefficient, meaning that as of 2012, one would predict that Zambian firms were generally not equipped with generators.

3.2.8 *Extent of Use of Backup Generation*

Two relationships that Bental and Ravid (1982) surmised were that

1. as generators become more expensive, firms will purchase less backup power; and the damage of outages will increase; and
2. the more reliable the supply of on-grid electricity, the higher the cost of outages, as a result of decreased backup facilities purchased by the firm. Steinbuks and Foster (2010) also casually observed that the capital cost of generators tends to be higher in countries with more reliable energy.

Even with self-generated energy, Sichone et al. found that a percentage increase in use of self-generated electricity was associated with a 0.19% decrease in annual sales (Sichone et al., 2016). This implies that self-generation does not fully cover loss in energy.

Oseni and Pollitt posited and then showed that a firm that has invested in backup generation may still suffer a higher unmitigated outage loss relative to a non-backup firm if its operations are more vulnerable to power outages and its investment in self-generation capacity is not large enough to significantly reduce the potential outage loss (Oseni & Pollitt, 2015).

To test this theory on 2012 data collected by the World Bank with outage losses as the dependent variable, rather than using a pooled regression for firms that had generators and those that did not, Oseni and Pollitt ran separate regressions to account for the possibility that the generator variable may interact with other explanatory variables. The adjusted R-squares using this approach for the outage loss given backup generation or given no backup generation were 42% and 52% respectively for 1648 and 1017 observations across Africa.

A remarkable result that comes out of their regressions is that firms with backup generation were slightly more (4% more) vulnerable to outage losses than firms without backup generation capacity when the electricity consumption of the firm is greater.[5] This could possibly mean that there are two types of large consumers of energy: those that need to mitigate the losses with backup generation but cannot afford to mitigate the loss entirely, and there are those for which the costs of backup generation do not equal the marginal benefit of self-generation. This may reflect the differences in the energy intensity of the two types of firms' operations, as Oseni and Pollitt note (2015).

An alternative explanation for this result could be that the underlying data is unsafe: it comes from questions asking the value and percentage of losses due to power outages. Responses could have been affected by the biases discussed above such as protests or loss aversion. The implication would be that respondents at firms with generators would protest more about their losses or feel their unmitigated losses even more than respondents at firms without generators. However, given that for the other non-country statistically significant variables the costs of outages were smaller for firms with generators than those without, this is unlikely to be the case. This was the expected result, and Oseni and Pollitt found that despite the substantial amounts of unmitigated outage losses suffered by firms with generators, their counterfactual estimates showed that firms engaging in self-generation were better off than if they did not invest in self-generation (Oseni & Pollitt, 2015).[6] In fact, their results showed that the actual unmitigated outage loss per kWh was considerably higher for backup firms than for non-backup firms in countries with frequent power outages.

In spite of the existing commentary on backup generation failing to fully make up for lost energy, research published as recently as 2018 in a *Nature* publication used the assumption that firms with generators used them to mitigate all outage losses (Farquharson et al., 2018).

[5] For the other non-country statistically significant variables for both firms that had generators and firms that did not have generators—days of power outages, employment, the chemical and pharmaceuticals subsector at the 10% level of significance—the outage losses were smaller for firms with generators than those without.

[6] They calculated the outage loss per kWh unsupplied assuming that a firm would have consumed the same amount of electricity per hour during outages as they did during uninterrupted hours, and assumed firms operated 12 hours per day.

3.3 The Impact of Power Outages in Zambia Prior to 2015

Before 2015, Zambia's power capacity theoretically should have allowed electricity supply to exceed industrial and mining demand, but news reports and multilateral surveys suggest that poor maintenance and corruption were responsible for power outages nonetheless.

While Zambia's electricity was amongst the cheapest in the world (Jourdan, 1990, p. 182), an accountant assigned to a consulting project in Zambia 1992 and 1993 recalls (email from Michael Mainelli dated 17 February 2019) that power quality was an issue in Zambia for private investors even then. 'One of the points we failed to get the government to take seriously enough', he writes,

> was that variability in the electricity supply was leading to production problems and thus variability in the output. In turn, these production problems were impeding the sales potential for Zambian copper as buyers were concerned about irregularity of supply. Clearly Zambia had significant hydropower production potential, but it was often out of action, or diverted elsewhere for political purposes [...] A small feedback problem related to this was that because Zambia's intermittent supply caused sub-optimal contracts, it in turn caused significant [foreign exchange] problems in turn, a second-order effect.

An example of a reason for the power stations being out of action was in 1989 when there was a fire at the Kafue station (Jourdan, 1990, p. 182). Power outages continued to be reported in the 2000s and mining companies (both open-pit and underground mines) responded by installing diesel generators to make up for the shortfall of delivered grid electricity to prevent ZESCO power outages from resulting in loss of production (Mfula, 2010a, 2010b; Steel News, 2011; Syndigate, 2011), although underground mines are more electricity intense (Tembo, 2018, p. 41).

Although mining is by far the largest of Zambia's electricity users, its energy consumption patterns do not appear to be diverse. Most of Zambia's mining is for copper, and 80% of Zambia's copper comes from just four mines (Zambia Chamber of Mines, 2016).[7] Line 3 of Table 3.1

[7] Copper and copper products accounted for 92% of Zambia's exported metals by value in 2018 and 72% of all exports (World Bank, 2020). The Lumwana and Kansanshi mines are open-pit (Barrick Gold Corporation, 2019; First Quantum Minerals Ltd, 2019), whereas

Table 3.1 Zambian copper mining production 2013–2018e

Imputing diesel-generated electricity by the mining sector

Line	Year	2013	2014	2015	2016	2017	2018
1	Mining grid electrcity consumption, GWh	5929	5871	6246	5918	6202	6682
2	Copper (Cu) production, '000s tonnes	760	708	712	763	794	854
3	Grid electricity/Cu produced, GWh/'000 tonnes	7.8	8.3	8.8	7.8	7.8	7.8
4	Median average electricity/Cu producted, GWh/'000 tonnes	7.8					
5	Predicted grid electricity consumption given Cu produced, GWh	5941	5535	5566	5965	6207	6676
6	Variance where predicted grid electricity consumption > actual grid electricity consumed, GWh	12			47	5	

Sources: Energy Regulation Board of Zambia, 2015, p. 8, 2016a, p. 9, 2017a, p. 9, 2018, p. 36, 2019, p. 39 and U.S. Geological Survey, 2015, 2016, 2017, 2018, 2019, 2020 and own computations

shows that for each of 2013, 2016, 2017 and 2018, the mining grid electricity consumption in gigawatt-hours per copper production in thousands of tonnes ratio was the same to two significant figures (Figs. 2.6 and 2.7).

Using this ratio, we can calculate how much grid electricity consumption mining would have required where we have the copper output figures in Line 5 of Table 3.1. Where this calculated figure exceeds the Energy Regulation Board's reported figure of energy consumption by the mining sector, the difference is what we can estimate as the amount of energy mining companies self-generated, so 12 GWh in 2013, 47 GWh in 2016 and 5 GWh in 2017. Note that the prediction shows that in 2015, they did not have to self-generate. The mining sector was somewhat more insulated in the beginning (ZESCO Ltd, 2017a). Indeed, in 2015, mining's share of ZESCO energy grew from 47% in 2014 to 55%. But then ZESCO requested the mining sector to reduce its load by 30% (Energy Regulation Board of Zambia, 2016a, p. 1), and mining consumption data of ZESCO energy indicate that the period of worst load shedding for the mining sector was from December 2015 to November 2016.

In spite of that, Zambia's copper mining production did not fall either from 2014 to 2015 or from 2015 to 2016 (U.S. Geological Survey, 2015, 2016, 2017, 2018, 2019), but instead grew in 2016 to its four-year peak (see Table 3.1). That copper production grew in spite of heavy power outages of December 2015 to November 2016 implies that it was still economic to extract copper and supports Tembo's findings that the key decision variables for Zambia's mining sector are copper price, grade and type of ore, and not energy price (Tembo, 2018). Given that 60% of

Mopani and Konkola Copper are a mix of underground and open-pit mines (Konkola Copper Mines Plc, 2019; ZCCM Investment Holdings Plc, 2019).

Zambia's mines are open-pit (Tembo, 2018, p. 50) and less electricity intense, it is probable that the proportion of production from less electricity-intense open-pit mines increased.

Moving on to manufacturing, Steinbuks and Foster (2010) provide one of the earlier studies on the impact of power outages on Zambia's manufacturing sector using the World Bank's Enterprise Survey data collected from 2002 to 2006. Providing their assumed cost of self-generating electricity (variable and capital charge for generators) and the grid cost of electricity, they provided their output average cost of energy. Working their calculation backwards (Steinbuks & Foster, 2010, Table 2.1),[8] we find that Zambian manufacturing firms experienced power outages for 5% of their operating hours at the time that the data that they used was calculated.

They also found that firms experienced a 7.3× difference in losses between having a generator and not having a generator (Steinbuks & Foster, 2010, Table 2.2). The value they used to ascribed to having a generator was the capital and variable cost of a generator. The weakness associated with how they calculated the capital cost has been discussed above. The value they ascribed to not having a generator was the average loss of sales over the course of an hour. The weaknesses with this are that manufacturing is a heterogeneous sector, with diverse subsectors, and that firms tend to overstate their losses, as noted above. Nonetheless, because power outages occurred with such infrequency, they found that the costs of having a generator exceeded the benefits (Steinbuks & Foster, 2010, Table 5a), and that only 38% of firms owned generators (Steinbuks & Foster, 2010, Table 1.1).

Further research by Sichone et al., discussed above, used the World Bank's Enterprise Surveys of 2013.

The critical juncture came in 2015, however, as Fig. 1.1 illustrates, when power outages resulted systematically and frequently from severe droughts, causing growth of manufacturing to decline The drop in growth of the manufacturing was followed by a steep decline in net inflows of foreign direct investment in 2016, as illustrated in Fig. 2.6.

[8] Using the Goal Seek function in MS Excel for USD 0.45/kWh × x% of the time + USD 0.04/kWh × (1−x%) of the time = USD 0.06/kWh.

3.4 The Impact of the 2015 and 2016 Power Outages in Zambia

Headline statistics suggest that the power outages of 2015 and 2016 adversely affected manufacturing firms, and annual reports from listed companies confirm this impression. Between the power outages of 2015 and 2016, Zambian manufacturing's share of electricity consumption fell from third biggest sectoral consumer to fourth biggest sectoral consumer behind finance and property (see Table 1.1), and manufacturing value-added growth fell (Fig. 1.1). Table 3.2 shows how the annual reports of eight listed Zambian manufacturing firms reported the impact of power outages on their operations. Six out of seven firms whose annual reports for 2016 were obtained mentioned the power outages, and of them, four mentioned them in the context of adversely affecting firm performance or production processes.

The power outages seem to have affected firms differently depending on their subsector. Cement manufacturer LaFarge experienced reduced revenues and profits year on year from 2014 to 2016, attributing the decline in performance to power shortages which adversely impacted production. In contrast, Zambia Sugar Plc saw increased revenues year on year from 2014 to 2016. Because Zambia Sugar Plc was sourcing energy from the fibrous residue remaining after the extraction of sucrose from sugar cane, it was not heavily dependent on ZESCO for energy: ZESCO energy was a supplementary source during the factory off-crop maintenance period when the factory was not operational (Zambia Sugar, 2017, p. 15). The impact of power outages for this agricultural firm, therefore, were negative primarily through the lack of irrigation, rather than through the impact on the firm's operating factory.

'PF' in Table 3.2 indicates a

- direct attribution of decline in company performance from the previous year due to power outages:
  - for example, LaFarge Zambia reported in its 2015 annual report 'profit before tax was 25% down from 2014 due to fast rising costs, particularly power and major input costs' (Lafarge Zambia, 2016).
- decline in performance in terms of where the company would otherwise have been:

Table 3.2 Annual reports of listed Zambian manufacturers mentioning the power outages

Key:

Reduced revenues and profits from previous year
Increased revenues but reduced profits from previous year
Reduced revenues but increased profits from previous year
Increased revenues and increased profits from previous year

PF – Power shortages mentioned as a cause of decline of firm performance
PP – Power shortages mentioned as adversely impacting production
PE – Power shortages mentioned as a cause of slow-down of economic growth
T – Mention electricity tariff increase as adversely affecting the company
IF – Improved performance of firm from better power delivery
IP – Improved processes as a result of better power delivery
IE – Improved economic performances from previous year from better power delivery

Firm & annual reports referenced	2015	2016	2017
LaFarge (2016, pp. 11–12, 17, 2017, pp. 3–4, 36, 43, 2018, pp. 4–5, 50, 60)	PF PE	PF PE	T
British American Tobacco (2016, pp. 3–4, 6, 2017, pp. 9, 14, 2018, pp. 8, 23)	PE		IE
National Breweries (2016, pp. 18, 24, 2018, pp. 4, 27)	PE	PE	IP
Zambia Bata Shoe Co (2016, p. 10, 2017, pp. 4, 14, 2018)		PP, PF	
Zambeef (2016, pp. 3, 10, 18, 45, 2017, pp. 23, 45, 2018, pp. 2, 10–11, 14–15, 17–18, 53)	PP PE	PP	T IP
Metal Fabricators of Zambia (ZAMEFA, 2016b, p. 5, 2016a, pp. 1, 4, 10, 2018, p. 7)	T	PE	
Zambia Sugar (2016, pp. 1, 11–12, 20, 2017, pp. 24, 41, 2018, pp. 8, 31)	PP Reduces demand for on-grid electricity by generating power for its factory using its waste biomass.	PP Ability to produce own electricity has minimised the impact of load shedding on our operations.	IP
Zambia Breweries (National Breweries, 2018)			IP IE
Analysis			
Power outages mentioned	5/7	6/7	
Power outages mentioned as adversely affecting firm performance or production processes (PF, PP)	3/7	4/7	
Improvement in on-grid power mentioned			5/7

Source: Ahmed et al. (2019)

– for example, Zambia Bata Shoe Company reported lower revenues but higher profits in 2016 from 2015, and attributed the lower sales to the higher costs of production due to power outages: 'We had also the load shedding due to lack of water in our major dams and rivers for generation of power. Our manufacturing had to run on generators and this increased our input costs reducing our margins as we could not afford to increase our products since most of our customers could not afford such price increases' (Zambia Bata, 2017).

While power shortages were one key feature of 2015 and 2016 that adversely affected their operations, the annual reports for the food processing companies also mentioned the drought (which also caused the power shortages), increases in fuel costs (which would have been caused by higher demand for diesel fuel for self-generation), inflation (which would have been caused by increased fuel costs), depreciation of the local currency (which would have been caused by inflation), high interest rates (which would have been imposed to control inflation) and low copper prices (likely determined by exogenous market forces). There were other factors besides power outages that were contributing to the fall of manufacturing value-added.

For a further idea of how the power outages impacted businesses in Zambia in 2015 and 2016, we need to broaden our review of the literature to research carried out by Mwila et al. (2017) who surveyed the impact of power outages on micro-, small- and medium-sized businesses, which focused mainly on services since the percentage of the MSME population engaged in manufacturing was only 9.3% (Mwila et al., 2017, Fig. 4.13).[9] Similarly, 93% of Batidzirai et al. (2018)'s

[9] According to the MSME policy of the Ministry of Commerce, Trade and Industry (Zambia Development Agency, 2019), micro, small and medium enterprises are those registered with the Patents and Companies Registration Agency (PACRA) with the following characteristics (Kwacha converted to US dollar at a rate of 13 per USD as of 27 September 2019).

Mwila et al. estimated that small enterprises across sectors accounted for over 170,000 jobs, although the methodology used to estimate this number was not revealed. Half of small businesses were involved in trade (2017, p. vi). Of MSMEs falling within the manufacturing sector, they only contributed to 1.5% the manufacturing sector's contribution to GDP (Ministry of Commerce Trade and Industry, 2014, p. x), even though small and medium businesses accounted for 56% of the manufacturing sector population in 2011–2012 (Ministry of Commerce Trade and Industry, 2014, p. x).

sample was micro and small enterprises. The one finding they found germane to large-scale manufacturing was that a manufacturing firm with 400 employees was willing to pay ZMK 0.32 more for reliable electricity, although whether this was per kWh is not clear (Batidzirai et al., 2018, p. 18). They also found that overall manufacturing firms were more willing to pay for reliable electricity services than other sectors, followed by financial services (Table 3.3).

Small businesses were less resilient than larger firms because of limited resources to invest in alternative energy sources and most lacked insurance (Mwila et al., 2017, p. v).

Drawing a sample of 696 small enterprises that they interviewed in 2015 from an estimated population of 15,415, Mwila et al. found the following:

- Electricity featured as the most popularly reported operational constraint to firms, being voted as the top constraint by 36% of respondents. Access to finance ranked second with 28% of votes and competition ranked third with 20% of votes. These were followed by 'other', labour, security and fuel (Mwila et al., 2017, p. 20).

  - It should be noted that electricity supply moved up from third most popularly voted operational constraint after access to finance and the informal sector in the World Bank's Enterprise Survey (World Bank, 2014, p. 4) conducted before the outages. The enterprise surveys excluded companies engaged in copper mining. 51% of surveyed firms were in manufacturing.
  - By contrast, however, electricity supply deficit was ranked third most popular constraint in the Bank of Zambia, Central Statistical Office and Zambia Development Agency's 2016 survey (Bank of

Table 3.3 How the Zambia Development Agency categorised micro-, small- and medium-sized companies

Category	*Micro enterprise*	*Small enterprise*	*Medium enterprise*
Investment in equipment, $	77–6153	6154–15,384	15,385–38,461
Annual sales, $	77–11,538	11,539–23,076	23,077–61,538
Workers	≤ 10	11–50	51–100

Sources: Zambia Development Agency, 2019; Foreign exchange bureaus in Lusaka

Zambia, Central Statistical Office and Zambia Development Agency, 2017, p. x) of 266 firms across sectors, after exchange rate instability and the high cost of borrowing. The difference could have been that Mwila et al. focused on small businesses whereas the latter also included the mining sector. It seems that the ranking of the importance of electricity outages in the three sets of surveys was dependent on the time of survey, the sample of respondents, and the options respondents were given. Given the different results of the Bank of Zambia's survey conducted at the time of the worst power outages, it is difficult to conclude that power outages were the biggest challenge to firms, but we can conclude that they were an important challenge.

- In terms of average load-shedding hours per day, Lusaka was hit the worst (Mwila et al., 2017, p. 24), but in terms of the cost per kilowatt hour cost, Kitwe was had the highest loss (Mwila et al., 2017, p. viii).
- While only 17% of SMEs reported not having received notifications of ZESCO's load shedding schedule through any media or members of the public (Mwila et al., 2017, p. 25), 51% of enterprises reported that ZESCO did not follow its own load shedding schedules (Mwila et al., 2017, p. vii).
- A total of 30% of businesses reported damaged equipment due to load shedding. Only 12% of businesses reported their equipment as insured. Firms also suffered from restarting operations costs (2017, p. vii, 30).
- Load shedding resulted in idle labour and overtime labour costs (p. 28). About 8% of firms reported reducing on labour hours (p. 36).
- Load shedding resulted in purchases and hiring costs for alternative energy sources (p. 31). Among the surviving firms that were interviewed, 42% reported to have used generators in 2015 (p. 33), whereas just 2% reported having used a UPS (p. 34) and just 2% reported having used surge protectors (p. 34).
- Only 0.2% of businesses reported moving (p. 32). However, the total number that did was 21 businesses, all in Lusaka (p. 37).
- Some 7% of businesses reported switching working hours (p. 38). The report also says that no businesses reported shutting down their operations (p. 37), but this result would be heavily affected by survival bias.
- Around 7% of small- and medium-sized businesses reported backing up data systems (p. 34).

- A total of 7% of sampled firms used enhanced security (p. 35), while 21% of firms reported facing theft during load shedding (p. 39).

Batidzirai et al. (2018) surveyed 224 mostly micro and small enterprises (this constituted 93% of their sample, pp. 8, 14) on their willingness to pay a higher electricity tariff for reliable service. They concluded that an upward revision of electricity tariffs would facilitate investment in the energy sector, and that most businesses would be willing to pay for improved electricity supplies. Indeed, Zambia's average electricity tariff was as of June 2016 third lowest out of ten in the Southern African Development Community, and less than half of Namibia's (Energy Regulation Board of Zambia, 2017, p. 61).

They found that hotels, restaurants and businesses paying a commercial tariff (i.e. other service sector companies) as well as operating hours per day were statistically significant independent variables in a regression where willingness to pay more was the dependent variable (Batidzirai et al., 2018, p. 23). Variables that did not seem to have an impact were the number of years a firm had been in business, whether it was profitable and whether a business was a service sector business that was not hospitality (they looked at the financial and wholesale and retail subsectors).

However, it is unclear when their 224 surveys were conducted. They cite electricity tariffs approved by the Energy Regulation Board in March 2017. Because ZESCO made tariff revisions in May 2017 and then again in September 2017, it is unclear whether Batidzirai et al.'s findings were relevant even at the time of publication (Table 3.4).

While finance and property seem to have fared better through the power outages of 2015 and 2016 than manufacturing firms as suggested by their increase in share of energy consumption, the services sector in general and SMEs did rank power outages as the most popular constraint to their operations.

3.5 The Impact of Power Outages on the Climate

Beyond looking at the impact of power outages on ZESCO's largest customers, it is relevant to consider their climate impact, since global warming is what caused the power outages in the first place. Farquharson et al. (Farquharson et al., 2018) based their top-down analysis of generator emissions for Zambia, at least, using out of date because that was collected by the World Bank from December 2012 to February 2014 (World Bank

Table 3.4 ZESCO saw several variable energy tariff revisions 2017–2020

Customer category		*Tariffs prior to 15 May 2017*	*New Tariff 15 May 2017*	*New Tariff 1 Sept 2017*	*New Tariff 1 Jan 2020*
1. Metered Residential Tariffs (Prepaid) (Capacity 15 kVA)					
R1-Consumption ≤200 kWh/month	Energy Charge/kWh	0.15	0.15	0.15	
R2-Consumption >200 kWh/month	Energy Charge/kWh	0.51	0.77	0.89	
R1 revised as of 1 Jan 2020: ≤100 kWh/month					0.47
R2 revised as of 1 Jan 2020: 101–300 kWh/month					0.85
R3 revised as of 1 Jan 2020: >300 kWh/month					1.94
2. Commercial Tariffs (Capacity 15 kVA)					
Commercial	Energy Charge/kWh	0.31	0.47	0.54	
C1 as of 1 Jan 2020: ≤200 kWh/month					1.07
>200 kWh/month					1.85
3. Social Services Tariffs					
Schools, Hospitals, Orphanages, Churches,	Energy Charge/kWh	0.28	0.42	0.49	1.19
4. Maximum Demand Tariffs					
MD1-Capacity between 16 kVA and 300 kVA					
	Energy Charge(K/kWh)	0.2	0.3	0.35	0.61
	Off-Peak Energy Charge(K/kWh)	0.15	0.23	0.26	0.46
	Peak Energy Charge(K/kWh)	0.25	0.38	0.44	0.77
MD2-Capacity between 301 kVA and 2000 kVA					
	Energy Charge(K/kWh)	0.17	0.26	0.30	0.53

(*continued*)

Table 3.4 (continued)

Customer category		*Tariffs prior to 15 May 2017*	*New Tariff 15 May 2017*	*New Tariff 1 Sept 2017*	*New Tariff 1 Jan 2020*
	Off-Peak Energy Charge(K/kWh)	0.13	0.2	0.23	0.39
	Peak Energy Charge(K/kWh)	0.21	0.32	0.37	0.66
MD3-Capacity between 2001 kVA and 7500 kVA					
	Energy Charge(K/kWh)	0.14	0.21	0.25	0.43
	Off-Peak Energy Charge(K/kWh)	0.1	0.15	0.18	0.32
	Peak Energy Charge(K/kWh)	0.17	0.26	0.3	0.54
MD4-Capacity above 7500 kVA					
	Energy Charge(K/kWh)	0.12	0.18	0.21	0.36
	Off-Peak Energy Charge(K/kWh)	0.09	0.14	0.16	0.27
	Peak Energy Charge(K/kWh)	0.14	0.21	0.25	0.45

Tariffs are exclusive of 3% government excise duty and of 16% value-added tax

Sources: Author's adaption of ZESCO (2017b) and Energy Regulation Board (2019) source information

and International Finance Corporation, 2014), that is, before the power outages of 2015 and 2016 and before, as the next chapter shows, most manufacturing firms purchased their first diesel generators.

For Zambia, they estimated that a typical month saw 5.2 outages for an average of 2.8 months; that the installed grid capacity was 2.3 GW and that generator availability was 3% of the grid. This would imply that they estimated that generators on average produced 1 GWh of energy in a month, the equivalent of 3.6 TJ, resulting in just 74 tonnes of CO_2 in a month.

Their analysis also ignored the possibility that backup generation capacity may not be used at all times of power outages.

In the absence of data for load lost and replaced by backup generation (even proportion of generator ownership), Farquharson et al. used Monte Carlo Analysis in which the load replaced during outages was bounded by a distribution of backup capacity available. This was judged as 1–25% in the case of Zambia. Explanation was not provided for why 1%. The upper bound was the maximum observed regional value as a percentage of grid capacity. The region used for Zambia is not provided, nor the source of the value 25%. Farquharson claim that the mode was the current estimate of the installed backup capacity 3%, but do not explain mode of what. That their estimate in 2018 is therefore made using outdated data and that their calculations lack transparency leaves the impression that their findings on the estimated annual electricity replaced by diesel backup generators, and subsequent findings reliant on that estimate, are frail.

Addressing this issue with a top-down estimate, Ahmed et al. took a bottom-up sectoral approach to estimate the emission from backup diesel generator use by ZESCO, the mining sector, manufacturing sector and households. They estimated the emissions from power diesel generation use by ZESCO using the Energy Regulation Board's reports of diesel generation use. They estimated diesel generation for the relatively homogenous mining sector using secondary data on mining output, data on grid electricity consumed and assuming a constant ratio of electricity intensity for output to impute electricity that would have been generated by backup diesel generators. They estimated diesel generation use for manufacturing firms and households using primary collected survey data. In the months of worst outages for 2019, they estimated emissions to be north of 27,000 tonnes of carbon dioxide per month (Ahmed et al., 2020).

Ahmed et al.'s estimate is not of the same order of magnitude of Farquharson et al.'s estimate. To triangulate their results, they took 2018 as a base year of electricity consumption, since 2018 was not according to their household survey a bad year for power outages. Given that in the worst months of 2019, outages were lasting 15 hours, they estimated that if mining, industry and households generated 15% of grid capacity that they would have consumed in 2018, CO_2 emissions would have been 22,000 tonnes, so of the same order of magnitude as their bottom-up estimate.

Beyond estimating the volume of the greenhouse gas CO_2 emitted by diesel backup generators, Farquharson et al. also estimated the generated volume fine particulate matter, carbon monoxide, sulphur oxides and nitrogen oxides. Less important than the estimates that they calculated (founded on frail assumptions) is the attention that they brought to the fact that diesel backup generators increase greenhouse gases in the atmosphere and adversely impact health.

References

Adenikinju, A. F. (2003). Electric infrastructure failures in Nigeria: A survey-based analysis of the costs and adjustment responses. *Energy Policy, 31*(14), 1519–1530. https://doi.org/10.1016/S0301-4215(02)00208-2

Ahmed, I. (2021). *The political economy of energy mix in hydropower dependent developing nations–a case study of Zambia*. University College London. https://discovery.ucl.ac.uk/id/eprint/10121236/.

Ahmed, I., Baddeley, M., Coffman, D., Meikle, J., Oseni, M., & Sianjase, G. (2019). *The cost of power outages to Zambia's manufacturing sector*. https://www.theigc.org/wp-content/uploads/2019/08/Ahmed-et-al-2019-policy-brief-41408.pdf. https://doi.org/F-41408-ZMB-1

Ahmed, I., Parikh, P., Sianjase, G., & Coffman, D. (2020, December 4). The impact decades long dependence on hydropower in El Niño impact-prone Zambia is having on carbon emissions through diesel backup generation. *Environmental Research Letters, 15*(12) https://doi.org/10.1088/1748-9326/abb6a1

Allcott, H., Collard-Wexler, A., & O'Connell, S. D. (2015). *How do electricity shortages affect industry? Evidence from India, NBER Working Paper Series* 19977, Cambridge, MA. https://doi.org/10.1257/aer.20140389

Andersen, T. B., & Dalgaard, C. J. (2013). Power outages and economic growth in Africa. *Energy Economics, 38*, 19–23. https://doi.org/10.1016/j.eneco.2013.02.016

Arnold, J. M., Mattoo, A., & Narciso, G. (2008). Services inputs and firm productivity in Sub-Saharan Africa: Evidence from firm-level data. *Journal of African Economies, 17*(4), 578–599. https://doi.org/10.1093/jae/ejm042

Baddeley, M. (2003). *Investment: Theories and analysis*. Palgrave Macmillan.

Baddeley, M. (2013). Behavioural economics and finance. First edit. Routledge.

Bank of Zambia, Central Statistical Office and Zambia Development Agency (2017). *Foreign Private Investment & Investor Perceptions in Zambia—2017: Enhancing Investment for Export Promotion towards Inclusive Growth*. Lusaka: Government of the Republic of Zambia.

Barrick Gold Corporation. (2019). *Lumwana, Barrick website*. Retrieved October 2, 2019, from https://www.barrick.com/operations/lumwana/default.aspx

Batidzirai, B., Moyo, A., & Kapembwa, M. (2018). *Willingness to pay for improved electricity supply reliability in Zambia—A survey of urban enterprises in Lusaka and Kitwe*. IGC.

Beenstock, M., Goldin, E., & Haitovsky, Y. (1997). The cost of power outages in the business and public sectors in Israel: Revealed preference vs. Subjective valuation. *The Energy Journal, 18*(2), 39–61. https://www.jstor.org/stable/41322730

Bental, B., & Ravid, A. S. (1982). A simple method for evaluating the marginal cost of unsupplied electricity. *The Bell Journal of Economics, 13*(1), 249–253. http://www.jstor.org/stable/3003447

Chen, C.-Y., & Vella, A. (1994). Estimating the economic costs of electricity shortages using input-output analysis: the case of Taiwan. *Applied Economics, 26*(11), 1061–1069. https://doi.org/10.1080/00036849400000122

Diboma, B. S., & Tamo Tatietse, T. (2013). Power interruption costs to industries in Cameroon. *Energy Policy, 62*, 582–592. https://doi.org/10.1016/j.enpol.2013.07.014

Eberhard, A., et al. (2008, May). Underpowered: The State of the Power Sector in Sub-Saharan Africa. *Background Paper 6, Africa Infrastructure Country Diagnostic*, p. 54. https://doi.org/10.1109/TENCON.2008.4766817

Energy Regulation Board of Zambia. (2015). *Energy Regulation 2014*.

Energy Regulation Board of Zambia (2016a). *Energy Sector Report 2015*. Lusaka. Available at: http://www.erb.org.zm/reports/EnergySectorReport2014.pdf.

Energy Regulation Board of Zambia (2017). *Energy Sector Report 2016*. Lusaka. https://doi.org/10.13140/RG.2.1.3893.7043.

Energy Regulation Board of Zambia. (2018). *Energy Sector Report 2017*. http://www.erb.org.zm/downloads/reports/esr2017.pdf

Energy Regulation Board of Zambia. (2019). Public notice—Board's decision on ZESCO's application to adjust electricity tariffs and connection charges for the year 2019.

Escribano, A., Guasch, J. L., & Pena, J. (2010, January). Assessing the impact of infrastructure quality on firm productivity in Africa: Cross-country comparisons based on investment climate surveys from 1999 to 2005, p. 117. https://doi.org/10.1596/1813-9450-5191

Farquharson, D., Jaramillo, P., and Samaras, C. (2018, October). Sustainability implications of electricity outages in sub-Saharan Africa', Nature Sustainability, 1, 589–597. https://doi.org/10.1038/s41893-018-0151-8 Sustainability.

First Quantum Minerals Ltd. (2019). *Kansanshi, first quantum minerals website*. Retrieved October 2, 2019, from https://www.first-quantum.com/Our-Business/operating-mines/Kansanshi/

Fisher-Vanden, K., Mansur, E. T., & Wang, Q. J. (2015). Electricity shortages and firm productivity: Evidence from China's industrial firms. *Journal of*

Development Economics, 114, 172–188. https://doi.org/10.1016/j.jdeveco.2015.01.002

Foster, V., & Steinbuks, J. (2009). *Paying the price for unreliable power supplies—In-house generation of electricity by firms in Africa, The World Bank Policy Research Working Paper*. https://doi.org/10.1596/1813-9450-4913

IMF. (2008). *Regional economic outlook—Sub-Saharan Africa*. International Monetary Fund.

Jaramillo, P., & Skoknic, E (1973). *Costo Social de las Restricciones de Energia Electrica*. Oficina de Planificacion, Chile, August.

Jourdan, P. P. (1990). *Strategies for the regional planning of the minerals industry in Southern Africa: The case of the SADCC*. University of Leeds. http://etheses.whiterose.ac.uk/490/1/uk_bl_ethos_291037.pdf

Kahneman, D., & Tversky, A. (1979). Prospect theory: An analysis of decision under risk. *Econometrica, 47*(2), 263–292. https://www.jstor.org/stable/1914185

Kessides, C. (1993). *The contributions of infrastructure to economic development—A review of experience and policy implications*. The World Bank.

Konkola Copper Mines Plc. (2019). *Corporate Profile, Konkola Copper Mines website*. Retrieved October 2, 2019, from http://kcm.co.zm/corporate-profile/

Lafarge Zambia. (2016). Annual Report 2015. Lusaka.

Mfula, C. (2010a). First Quantum's Zambia mine hit by power outage. *Reuters*. https://www.reuters.com/article/update-2-zambia-mining-power-idAFLDE6180X820100209

Mfula, C. (2010b, August 21). Singapore firm to spend $108 mln on Zambia coal mine. *Reuters*. https://uk.reuters.com/article/mining-zambia/singapore-firm-to-spend-108-mln-on-zambia-coal-mine-idUKLDE67K09320100821

Ministry of Commerce Trade and Industry. (2014). *Manufacturing Sector Study Report: 2011–2012*.

Mwila, A., Sinyenga, G., Buumba, S., Muyangwa, R., Mukelabai, N., Sikwanda, C., Chimbaka, B., Banda, G., Nkowani, C. and Bwalya, B., 2017. *Impact of load shedding on small scale enterprises*. [online] Lusaka: Energy Regulation Board. Available at: [Accessed 22 June 2021].

Naqvi, S. N., & Ahmed, A. M. (1986). *Preliminary revised PIDE macro-econometric model of Pakistan's economy*. Institution of Development Economics.

Olley, G. S., & Pakes, A. (1996). The dynamics of productivity in the telecommunications equipment industry. *Econometrica, 64*(6), 1263. https://doi.org/10.2307/2171831

Oseni, M. O., & Pollitt, M. G. (2015). A firm-level analysis of outage loss differentials and self-generation: Evidence from African business enterprises. *Energy Economics, 52*, 277–286. https://doi.org/10.1016/j.eneco.2015.11.008

Pasha, H. A., Ghaus, A., & Malik, S. (1989). The economic cost of power outages in the industrial sector of Pakistan. *Energy Economics, 11*(4), 301–318. https://doi.org/10.1016/0140-9883(89)90046-7

Plott, C. R., & Zeiler, K. (2005). The willingness to pay-willingness to accept gap, the 'endowment effect,' subject misconceptions, and procedures for eliciting valuations. *The American Economic Review, 95*(3), 530–545.

Reinikka, R., & Svensson, J. (2002). Coping with poor public capital. *Journal of Development Economics, 69*(1), 51–69. https://doi.org/10.1016/S0304-3878(02)00052-4

Samuelson, P. A. (1962). Parable and realism in capital theory: The surrogate production function. *The Review of Economic Studies, 29*(3), 193–206. https://doi.org/10.2307/2295954

Sanghvi, A. P. (1982). Economic costs of electricity supply interruptions. US and foreign experience. *Energy Economics, 4*(3), 180–198. https://doi.org/10.1016/0140-9883(82)90017-2

Sichone, Y. M., et al. (2016). Electricity load shedding. An econometric analysis of the productivity of firms in the manufacturing sector in Lusaka. *International Journal of Commerce and Management Research, 2*(12), 151–157.

Solow, R. M. (1956). A contribution to the theory of economic growth. *The Quarterly Journal of Economics, 70*(1), 65–94. http://www.jstor.org/stable/1884513

Steel News. (2011). Konkola Copper Mines to invest USD 33 million for diesel generators. *Steel Guru.* https://steelguru.com/steel/konkola-copper-mines-to-invest-usd-33-million-for-diesel-generators/242497

Steinbuks, J., & Foster, V. (2010). When do firms generate? Evidence on in-house electricity supply in Africa. *Energy Economics, 32*(3), 505–514. https://doi.org/10.1016/j.eneco.2009.10.012

Syndigate. (2011, June 17). KCM deploys 3 diesel generators at Konkola Deep Mining Project. *MENA Report.* https://search-proquest-com.libproxy.ucl.ac.uk/docview/872316305?rfr_id=info%3Axri%2Fsid%3Aprimo

Tembo, B. (2018). *Strategic investment decisions in Zambia's mining sector under a constrained energy system.* University College London. https://discovery.ucl.ac.uk/id/eprint/10048151/

Thaler, R. (1980). Toward a positive theory of consumer choice. *Journal of Economic Behavior and Organization, 1*, 39–60. https://doi.org/10.1016/0167-2681(80)90051-7

U.S. Geological Survey. (2015). *Mineral commodity summaries 2015, Mineral Commodity Summaries.* Reston, VI. https://s3-us-west-2.amazonaws.com/prd-wret/assets/palladium/production/mineral-pubs/copper/mcs-2015-coppe.pdf

U.S. Geological Survey. (2016). *Mineral commodity summaries 2016, Mineral Commodity Summaries.* Reston, VI. https://s3-us-west-2.amazonaws.com/

prd-wret/assets/palladium/production/mineral-pubs/copper/mcs-2016-coppe.pdf

U.S. Geological Survey. (2017). *Mineral commodity summaries 2017, Mineral Commodity Summaries*. Reston, VI. https://s3-us-west-2.amazonaws.com/prd-wret/assets/palladium/production/mineral-pubs/copper/mcs-2017-coppe.pdf

U.S. Geological Survey. (2018). *Mineral commodity summaries 2018, Mineral Commodity Summaries*. Reston, VI. https://s3-us-west-2.amazonaws.com/prd-wret/assets/palladium/production/mineral-pubs/copper/mcs-2018-coppe.pdf

U.S. Geological Survey. (2019). *Mineral Commodity Summaries 2019, Mineral Commodity Summaries*. Reston, VI. https://prd-wret.s3-us-west-2.amazonaws.com/assets/palladium/production/s3fs-public/atoms/files/mcs-2019-coppe.pdf

U.S. Geological Survey. (2020). *Mineral Commodity Summaries 2020*. Reston, VI. https://pubs.usgs.gov/periodicals/mcs2020/mcs2020-copper.pdf

Willis, K. G., & Garrod, G. D. (1997). Electricity supply reliability: Estimating the value of lost load. *Energy Policy, 25*(1), 97–103.

World Bank. (2018). *Data on infrastructure—World bank enterprise survey of business managers, enterprise surveys*. Retrieved November 28, 2018, from http://www.enterprisesurveys.org/data/exploretopics/infrastructure

World Bank. (2020). *Zambia Trade Summary 2018 Data, World Integrated Trade Solution*. Retrieved April 11, 2020, from https://wits.worldbank.org/CountryProfile/en/Country/ZMB/Year/LTST/Summary

World Bank and International Finance Corporation. (2014). *Zambia country profile 2013*. World Bank.

Zambia Bata. (2017). *Annual Report 2016*.

Zambia Chamber of Mines. (2016). *A concentrated mining sector, Mining for Zambia*. Retrieved October 1, 2019, from https://miningforzambia.com/a-concentrated-mining-sector/

Zambia Development Agency. (2019). *MSME Application Form*. Zambia Development Agency. http://www.zda.org.zm/?q=content/msme-appli-cation-form

Zambia Sugar. (2017). *2016 Annual Report*. https://africanfinancials.com/document/zm-zmsg-2016-ar-00/

ZCCM Investment Holdings Plc. (2019). *Mopani copper mines PLC, ZCCM-IH website*. Retrieved October 2, 2019, from https://zccm-ih.com.zm/copper-cobalt-gold/mopani-copper-mines/

Zesco Ltd. (2017a). *2016 Integrated Report*.

Zesco Ltd. (2017b). *Current Zesco tariffs, Zesco website*. Retrieved November 2018, from http://www.zesco.co.zm/customerCare/tariffs

CHAPTER 4

The Impact of the 2015 and 2016 Power Outages on Zambia's Manufacturing

Funding secured by the UCL Bartlett School of Sustainable Construction from the International Growth Centre enabled my field team led by my colleague Graham Sianjase and me to collect 123 surveys from large manufacturing firms stratified by sector and geographical location across Zambia from April to August 2018 to assess how power outages were impacting Zambia's manufacturing sector. This chapter shows the results of that fieldwork.

On a geographical basis, the aim was to have 59% of the sample come from Lusaka and the remaining from the Copperbelt (since 50% of large manufacturing firms are in Lusaka and 84% of large firms are in both Lusaka and the Copperbelt; 50/84.1 = 59%). In fact, 71% of the sample came from Lusaka, representing an overweight for Lusaka, even though the time enumerators in the Copperbelt were deployed was longer than the time they were deployed in Lusaka. The enumerators seemed to hit a limit as to how many firms they could interview in those two towns. This may or may not have been the result of fewer firms in the Copperbelt, particularly in Kitwe, surviving power outages.[1] The number of surveys in

[1] Kitwe seemed the hardest hit town of Lusaka, Ndola and Kitwe by power outages both according to Mwila et al. (2017) and according to the collected data. The Ministry of Trade and Industry's information may therefore have been more out of date for Kitwe than for other towns insofar as more firms may have gone out of business there because of the power outages.

I. Ahmed, *The Political Economy of Hydropower Dependant Nations*, https://doi.org/10.1007/978-3-030-71266-2_4

Table 4.1 Sampling achieved by subsector versus national population of large manufacturing firms by subsector

Large manufacturing firms by subsector and province

Subsector	Lusaka & Copperbelt	Central	Eastern	Luapula	Muchinga	Northern	NW	Southern	Western	Total	National	Achieved sample	Delta
Food and food products	96	7	6	2	1	7	3	15	2	139	36%	39%	-3%
Textiles and garments	15	0	2	0	0	0	0	1	0	18	5%	9%	-4%
Wood and wood products	21	0	0	0	0	0	1	2	1	25	6%	3%	3%
Chemicals	48	2	0	0	0	0	0	1	0	51	13%	16%	-3%
Plastics and rubber	25	1	0	0	0	0	0	0	0	26	7%	16%	-9%
Non-metallic mineral products	30	1	0	0	0	0	0	0	0	31	8%	1%	7%
Basic metals	19	2	0	0	0	0	0	1	0	22	6%	8%	-2%
Fabricated metal products	27	2	0	0	0	0	0	0	0	29	7%	12%	-5%
Machinery and equipment	28	0	0	0	0	0	2	0	0	30	8%	1%	7%
Electronics	10	0	0	0	0	0	0	0	0	10	3%	0%	3%
Other manufacturing	10	0	0	0	0	0	0	0	0	10	3%	0%	3%
Total	**329**	**15**	**8**	**2**	**1**	**7**	**6**	**20**	**3**	**391**			

Source: Author's calculation, Manufacturing survey; Ministry of Commerce Trade and Industry, 2014, Annex 7; Ahmed et al. (2020)

Ndola and Kitwe, the two major towns in the Copperbelt, was almost equal. Only one interview came from the seven manufacturing firms in the Lusaka South Multi Facility Economic Zone and only one firm in Kafue, in Lusaka Province.

On a subsector basis, no subsector was over or underrepresented by more than 9% (Table 4.1).

4.1 General Characteristics of the Manufacturing Sample

The geographical breakdown of the survey was as follows:

- 90% of 113 respondents were located in industrial zones or a Multi-Facility Economic Zone.
- 71% of respondents were based in Lusaka, 15% in Kitwe and 14% in Ndola.

The international characteristics of the sample were such that

- 35% of 123 respondents reported that they export.
- Of these, only four exported or had exported outside of Africa (QA3 of questionnaire in Appendix B). Three of these firms exported to China.[2] One firm exported to India previously but no longer exported anywhere. One firm exported to India in addition to China.

[2] Since this information is largely irrelevant to this study, this will not be discussed further, except to note that this tallies with the findings of Savin et al. (2020), who found that flows of official finance from China resulted in exports of manufactured goods from Africa to China.

- 58% of 42 questioned firms were majority foreign owned (QA5 of questionnaire in Appendix B).

Noting that the survey assessing the impact of power outages necessarily suffers from survival bias, the number of years of manufacturing experience of firms in the sample may be fewer than they would have been had there not been the power outages of 2015 and 2016.

Personnel at companies were asked for the number of years of that their firms had been manufacturing. Their responses were not counter-checked with official documents in most instances, so their responses should be taken as approximations and quoting the maximum number of years of experience may therefore be misleading. The median years of manufacturing experience was 15, and the interquartile range was 7–34. The least experienced firm whose staff were interviewed was less than a year old. The firm with the longest experience was older than Zambia having been formed before the nation's independence (Table 4.2).

A total of 14 of 119 companies (i.e. 12% of the sample) for which production hour data was collected manufactured 24 hours seven days of the week. However, the interquartile range of production hours was 48–89 hours. The distribution of production hours is represented by the below histogram (Fig. 4.1).

Table 4.2 Years of manufacturing experience in the sample of manufacturing firms

Measure	*Years of manufacturing experience*
Median	15
Standard deviation	19
Mean	22
Upper quartile	34
Lower quartile	7
Min	1
Max	Before independence

Source: Manufacturing survey. Author's calculation

Weekly hours worked	
119	Sample
168	Mode
75	Mean
48	Lower quartile
53	Median
89	Upper quartile

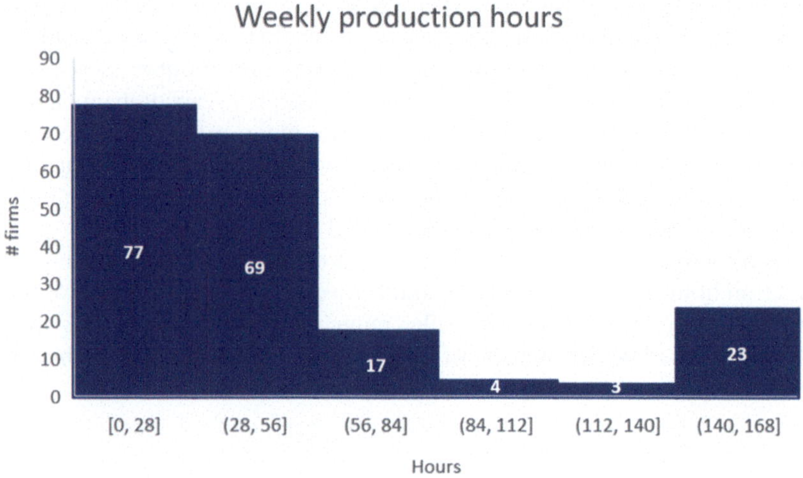

Fig. 4.1 Histogram and averages of production hours of the sample. (Source: Manufacturing survey)

4.2 Ranking the Damages Incurred to Large Manufacturing Firms from ZESCO's Outages

Respondents were asked to rank the various ways in which power outages imposed costs to them on a scale of 0–4, with 0 ranking as 'does not cause', 1 as 'causes to a minor degree', 2 as 'causes to a moderate degree', 3 as 'causes to a major degree' and 4 as 'causes all the time'.

With the exception of outages related to lapses in security, unplanned outages did more damage than planned damages for all types of damage recorded. However, 35 (24% of) firms surveyed interviewed said that they received no notifications of or unreliable notifications of ZESCO outages.

This meant that all outages for these firms were as bad as they could possibly be: they could not plan for outages.

Extra pay and transport costs for staff to stay on to work was the most common cost of power outages, followed by damage to equipment, damage to firm reputation and damage to inventory. The most serious cost—loss in clients—was not a common occurrence.

Figure 4.2 supports the work and findings of others (Pasha et al., 1989; Diboma & Tamo Tatietse, 2013) that planned outages considerably reduce costs when they come with notification. Unplanned outages resulted in additional costs due primarily to extra staff costs, secondly due to damage to equipment, thirdly due to damage to firm reputation and fourthly due to damage to inventory. By contrast, for planned outages, damage to equipment was on average considered to be 3.5× less costly, and damage to inventory was considered to be 4.7× less costly. For planned outages, extra staff costs again ranked higher, but so too did damage to firm reputation and loss in clients.

The only cost for planned outages which exceeded the same cost for unplanned outages was for security. This was presumably because it allowed firms to prevision for human personnel to secure premises when firms could expect security lights to go out.

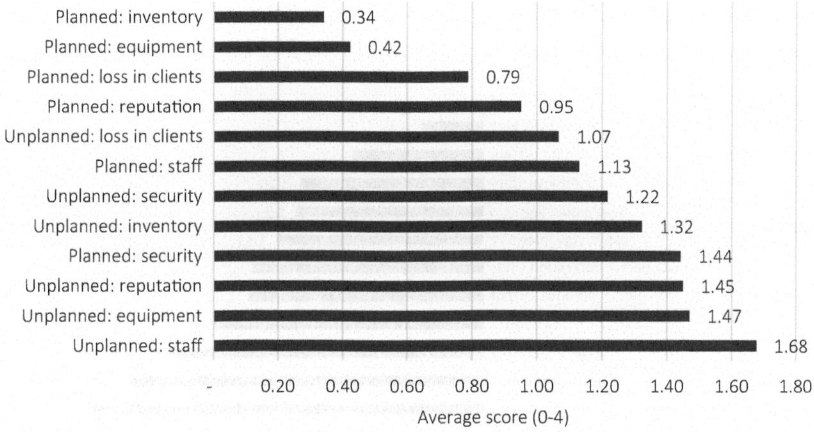

Fig. 4.2 Most common costs of power outages for manufacturing firms, April–August 2018. (Source: Manufacturing survey)

A quarter of ZESCO clients said that they never received notifications or accurate notifications for planned outages, meaning that in effect, ZESCO's failure to relay information accurately was costing its clients more. This negative experience would have informed their decision to say 'no' to paying ZESCO for more reliable energy.

4.3 Ranking the Use of Mitigation Strategies Used by Large Manufacturing Firms

Self-generation appeared as the third most popular mitigation intervention against power outages if the use of surge protectors, voltage regulators and capacitors were collapsed into one type of mitigation strategy.

Figure 4.3 tells us that interventions to mitigate damages to equipment and inventory are the most popular. Self-generation using backup diesel generators is next in popularity, followed by insurance against losses, switching labour to less energy intensive activities, rescheduling workers, and keeping larger inventories. Laying off workers was seldom a mitigation strategy that firms used.

Generator records showed that firms used their backup diesel generators more in 2015 and 2016 than in 2017 and 2018.

Generator records also showed that firms were still using backup diesel generation as recently as June, July and August 2018, which did not accord with what ZESCO's economists said (ZESCO economists, personal communication, June 6, 2018; Ahmed, 2021): that load shedding

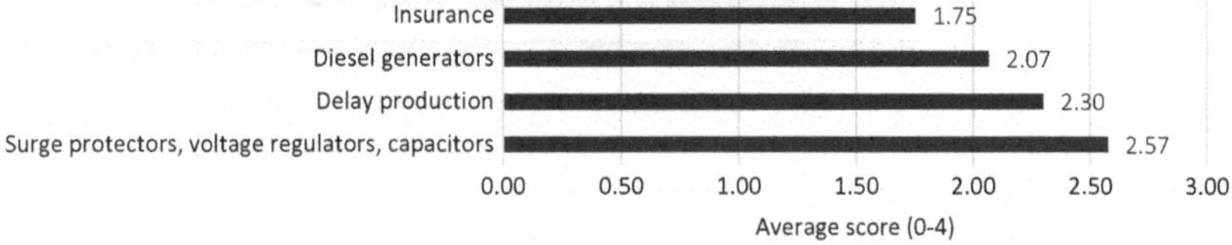

Fig. 4.3 Popularity of mitigation interventions for manufacturing firms, April–August 2018. (Source: Manufacturing survey)

came to an end in 2016 and that only maintenance outages take place on Sundays (it is possible that the power outages were not due to load shedding but other factors impacted by poor maintenance). ZESCO's own website shows load-shedding schedules past 2016, but shows none more recent than March 2017 (ZESCO Ltd., 2017b).

If data had been collected in 2019, it is thought self-generation patterns would have been closer to 2015 and 2016 patterns than to 2017 and 2018. Although 300 MW of coal energy had been added to the grid, most of ZESCO's power capacity continued to be vulnerable to low rainfall.

4.4 How the Costs of Power Outages Differ for Firms Using Differing Coping Strategies

In addition to the damages incurred from ZESCO's outages looked at in Sect. 4.2 and the use of mitigation strategies looked at in Sect. 4.3, this section looks at the association between the use of mitigation strategies and the extent of damages incurred. Descriptive statistics show that a greater extent of backup diesel generation use as a power outage mitigation strategy is associated with a reduction in rescheduling workers and with a reduction in loss of output. Econometric analysis shows that extent to which backup diesel generation is used and extent to which rescheduling workers are associated with a delay in production, and delays in production are associated with a loss in clients for both unplanned and planned outages. Delays in production and smaller inventories are associated with damage to firms' reputations for unplanned outages.

4.4.1 Models 1, 2 and 3

Models 1 and 2 shown in Table 4.3 that delay in manufacturing is positively correlated with loss in clients for both unplanned and planned outages.

Model 3 shows that an increase in the use of self-generation is associated with a decrease in delayed production and that an increase in rescheduling workers is associated with an increase in delayed production.

I tested the parallel regression assumption for validity for models 1–3, i.e. that all the coefficients on independent variables were equal for every ordinal value, using the Brant test. Brant tests could not be computed for Models 1 and 2 because there were not enough observations in some buckets. There were no observations where firms reported not delaying production at all AND reported loss of clients. This is consistent with the

Table 4.3 Ordinal logistic regressions for costs of power outages against mitigating strategies

Model	(1)	(2)	(3)	(4)
Consequences ?	Loss in clients		Use of delay	Damage to reputation
Coping mechanisms ?	Unplanned outage	Planned outage		Unplanned outages
Capacitors, voltage regulators, power factor correction units recat	0.309	0.381	0.134	0.178
	(0.218)	(0.283)	(0.233)	(0.206)
Delay	0.799***	1.145***		0.488**
	(0.275)	(0.389)		(0.248)
Self-generation	-0.299	-0.228	-1.053***	-0.162
	(0.227)	(0.290)	(0.250)	(0.218)
Larger inventory	0.106	0.218	0.290	-0.380*
	(0.227)	(0.283)	(0.258)	(0.223)
Resch workers	0.137	0.00800	0.727***	0.223
	(0.228)	(0.281)	(0.272)	(0.221)
u_inventory				
Constant cut3	3.283***	4.432***		1.800***
	(0.704)	(0.984)		(0.600)
Constant cut4	4.039***	5.220***		2.864***
	(0.746)	(1.053)		(0.647)
Constant cut1	1.163*	2.253**	-1.645***	-0.191
	(0.643)	(0.889)	(0.572)	(0.577)
Constant cut2	2.152***	3.634***	-0.485	0.988*
	(0.664)	(0.943)	(0.548)	(0.582)
Pseudo R-squared	7.8%	10.65%	16.64%	3.73%
Observations	115	81	117	114
Regression run	ologit	ologit	ologit	ologit
Standard errors in parentheses				
*** p<0.01, ** p<0.05, * p<0.1				
ologit = ordinal logistic				
Brant Test of Parallel Regression Assumption on variables, p>chi-squared values				
All	-	-	0.28	0.97
Capacitors, voltage regulators, power factor correction units recat	-	-	0.402	0.464
	-	-		
Delay recat	-	-	N/A	0.961
Self-generation recat	-	-	0.205	0.439
Larger inventory recat	-	-	0.316	0.898
Resch workers recat	-	-	0.312	0.749
Parallel assumption violated?			No	No

Source: Manufacturing survey. Author's calculation

Standard errors in parentheses; *** $p < 0.01$; ** $p < 0.05$; * $p < 0.1$

proposition that a delay in manufacturing results in loss in clients. The result was validated when allowing the effect to be different for each dependent variable category using a generalised ordinal logistic regression. Where the dependent variable of loss in clients was reduced further to a binary variable, so that 0 remained 0 and 1 and 2 became 1, the rescheduling workers had a statistically significant association with loss in clients for both unplanned and planned outages.

4.4.2 Model 4

With damage to firm reputation set as the dependent variable in model 4, several coping mechanisms appear to be associated. For unplanned outages, delays in production are positively associated and statistically significant at the 5% level. Larger inventories, a statistically significant coping mechanism at the 10% level for unplanned outages, seem to be an effective mitigation strategy. The chi-squared was not statistically significant for any of the tested variables in Model 4, which meant that the parallel regression assumption had not been violated for that model.

4.4.2.1 Discussion: Efficacy of Firm Interventions in Mitigating the Costs of Power Outages

The use of larger inventories was statistically significantly and negatively correlated with damage to firm reputation for unplanned outages.

The use of capacitors, voltage regulators and power factor correction units is statistically significantly positively correlated with damage to inventory for unplanned outages and damage to firm reputation for planned outages. However, this probably indicates reverse causality, as an alternative regression shows: damage to inventory is a positive and statistically significant predictor of use of capacitors, voltage regulators and power factor correction units. It is also useful to look at descriptive statistics to make sense of the results. About 20% of firms that had a voltage regulator, capacitor or power surge factor unit saw damage to their inventory or equipment, in contrast with 40% of firms that did not have these seeing damage to their inventory or equipment. Firms using these interventions therefore seem to have decreased their chances of damage by 50%. A follow-up study with semi-structured interviews could confirm the flow of cause and effect.

Self-generation marginally misses out being a statistically significant negative predictor at the 10% level for loss in clients for unplanned outages. It is, however, statistically significant at the 1% level as a negative predictor of delay, which in turn is a statistically significant predictor at the 1% level as a positive predictor of loss in clients for both unplanned and planned outages. Self-generation is also a negative predictor at the 10% level of statistical significance for damage to firm reputation resulting from planned outages. Although the flow of causality is a logical one and one would not expect reverse causality in these instances, to address criticisms of endogeneity through reverse-causality, a follow-up study with the same firms could clarify through semi-structured interviews what caused loss in clients, what caused production delays, how far self-generation mitigated these, and how decisions to invest in self-generation capacity and then use self-generation were made on an ex-ante and on an ex-post basis.

That more than 95% of firms that have access to a generator use their generators in spite of its high running cost of circa USD 0.29/kWh shows that they see value in using them. That there was a spike in acquisition of generators in 2008 suggests that there was latent demand for them prior to the duty waiver introduced in 2008 (Government of Zambia, 2008) for their import. They help mitigate against loss of welfare. From a public policy perspective, then, their import into the country should continue unimpeded.

4.5 Identifying Firm Characteristics That Can Be Used to Predict Differing Coping Strategies

Section 4.4 looked at the associations between coping strategies and costs of outages. This section looks at predictors of coping mechanisms.

4.5.1 Predictors of a Firm's Scale of Investment in Self-generation and Predictors of Ongoing Use of Backup Generation

4.5.1.1 Time as a Relevant Factor in the Purchase Decision—Descriptive Observations

Of 2014–2018, respondents at firms regarded 2016 as the worst year in terms of production losses, followed by 2015. The year 2017 was worse than 2018 (see Table 4.4 below).

Table 4.4 2016 saw even worse production losses than 2015, with surviving firms reporting losses on the higher spectrum of 16–30%. Even in 2018, firms were on average reporting production losses of 1–15%

Year	*2018*	*2017*	*2016*	*2015*	*2014*
Sum of 0–4 responses	47	66	92	73	40
Number of responses for year	42	42	38	32	25
Average score for year	1.12	1.57	2.42	2.28	1.60
Ranked in terms of worst year	5	4	1	2	3

Source: Manufacturing survey. Author's calculation

Surviving firms reported losses towards the higher spectrum of 16–30% in 2015 and 2016. Even in 2018, firms were on average reporting production losses of 1–15%.

2016 and 2015 ranked as worse years than the most recent years, meaning that even after discounting for what happened in the further past, respondents still rated these years as the worst for power outages.

We collected the purchase year of firms' oldest generators, a piece of datum that cannot be mistakenly reported by a respondent, and possibly a decent proxy for the year in which firms first bought generators. This rests on the assumption that generators are resilient and have long lifespans. Whereas three firms reported damaged generators that they had not replaced, six firms purchased their oldest generators in use before 2008, with the oldest generator in use having been purchased in 1994. A total of 36 out of 79 firms (45% of firms) bought their oldest running generators before the power outages of 2015. This would seem to support the notion that at least a large portion of generators still in use in 2018 were resilient.

The Zambian government waived duties on imported generators in 2008 (Government of Zambia, 2008). This could explain why 2008 is a modal year for the purchase of generators in the years up to 2013.

If the year of purchase of firms' oldest running generators is a good proxy for the year of purchase of firms' oldest generators, then the modal year for first-time purchases of generators was 2016. Of 79 responses, 21 (27%) of firms bought their first generator in 2016, while 20 (25%) of firms bought their first generator in 2015. Together, 2015 and 2016 accounted for 52% of first-time generator purchases (Fig. 4.4).

The correlation between the scores for how bad each year between 2014 and 2018 was shown in Table 4.4 and the number of generators that were purchased in 2014–2018 is 97%—see Fig. 4.5.

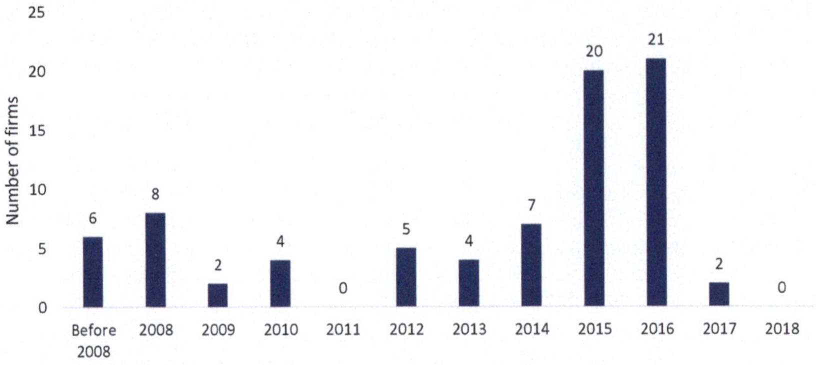

Fig. 4.4 Year in which first generator in use was purchased. (Source: Author's calculation, Manufacturing survey; Ahmed et al. (2020))

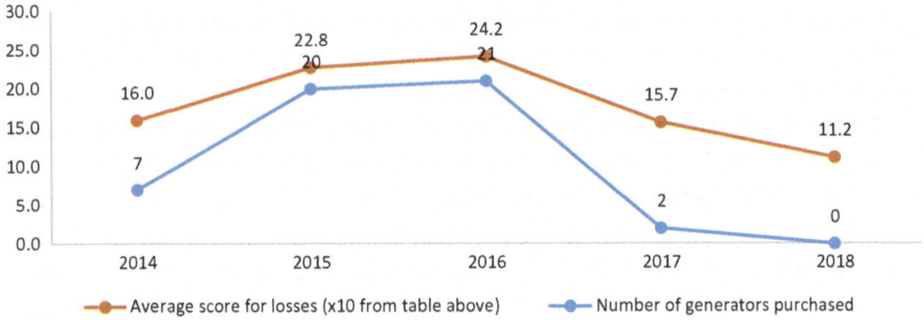

Fig. 4.5 Number of first generators bought by year; average score for losses by year (×10 from Table 4.6). (Source: Author's calculation, Manufacturing survey; Ahmed et al. (2020))

Worst Months Generally for Outages

Enumerators were asked to explore the seasonality of outages. This is relevant knowledge for manufacturing firms to be better prepared for when to expect power outages. Ninety-five votes were cast in response to the question of what months were worst for power outages by 31 distinct respondents. October was noted as the worst month for outages, followed by November and September. The month least reported for worst outages was April, which is in the dry season and when ZESCO produces less than average energy.

Discussion: How Worst Months for Power Outages Can Be Explained
Figure 4.6 illustrated how firms ranked months for worst power outages, reproduced as the navy columns in Fig. 4.7. Juxtaposed with average precipitation data as well as for energy consumption in 2014 to 2016, it becomes apparent why October ranks as the worst month: it follows months of no rainfall. Hydropower reservoir levels will be at their lowest.

Observed Generator Usage
Six large manufacturing firms shared their generation hours per month for varying periods. As the graphs show, the most concentrated period of outages was between June 2015 and October 2016. There was extreme use of self-generation in November and December 2017 for a particular company, and several companies were still using self-generation as late as April, May, June and July 2018 (Fig. 4.8).

In focusing on the hours of self-generation for the period January 2015–February 2017 for two large food and beverage companies, one located in Kitwe and one in Lusaka, it becomes apparent that the self-generation patterns are an almost perfect inflection of one another, suggesting that outages between the two cities could have been alternating (Fig. 4.9).

Diesel use by a large beverages company dramatically tailed-off in 2017 when reservoir levels returned to normal (Fig. 4.10).

4.5.1.2 Ownership Rate Versus Usage Rate—Descriptive Observations
Of 123 firms asked, 90 (i.e. 73%) responded that they had acquired use of a generator, through one mechanism or another (if not from outright purchase and sole ownership), and one firm responded that they were waiting for the shipment of a generator.

By contrast, 81 out of 122 firms (66%) said that they used their generator; that is, not all firms that had acquired access to a generator used them. The discrepancy for three firms that had acquired use of a generator at one time but did not use it was explained by the fact that their generators were no longer operational. One of them was engaged in trying to repair its generator. The other two did not report trying to repair or replace their generator. Taking these three firms out of consideration, 81 out of 87 firms (93% of firms) that had had access to a generator and whose generator did not break down continued to use their generator.

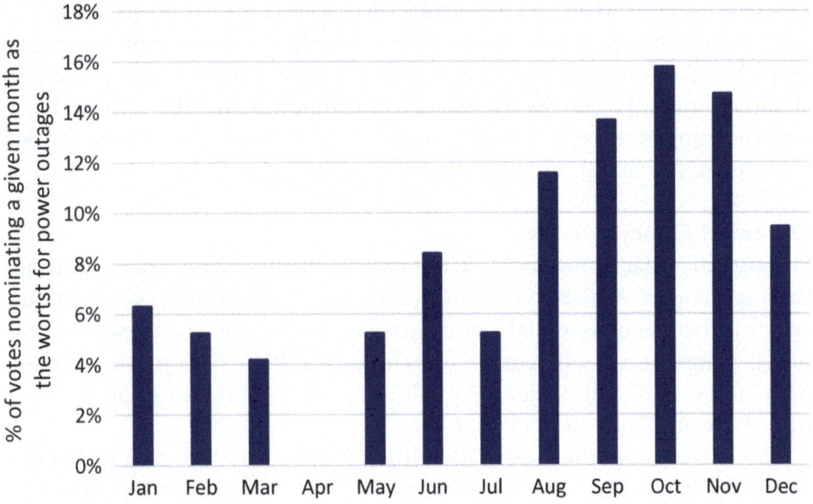

Fig. 4.6 Months rated as the worst for power outages. (Source: Manufacturing survey)

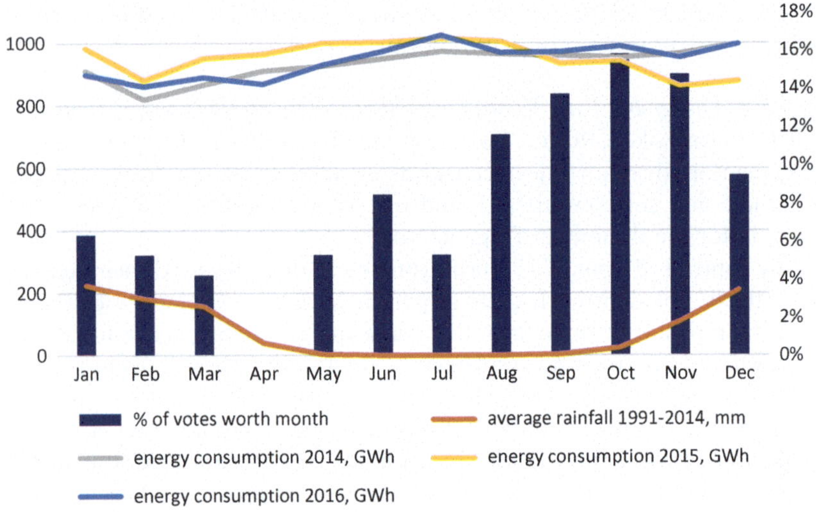

Fig. 4.7 October follows months of no rainfall is the worst month for power outages. (Author's creation. Sources: Manufacturing survey; ZESCO Ltd, 2017b; World Bank, 2018)

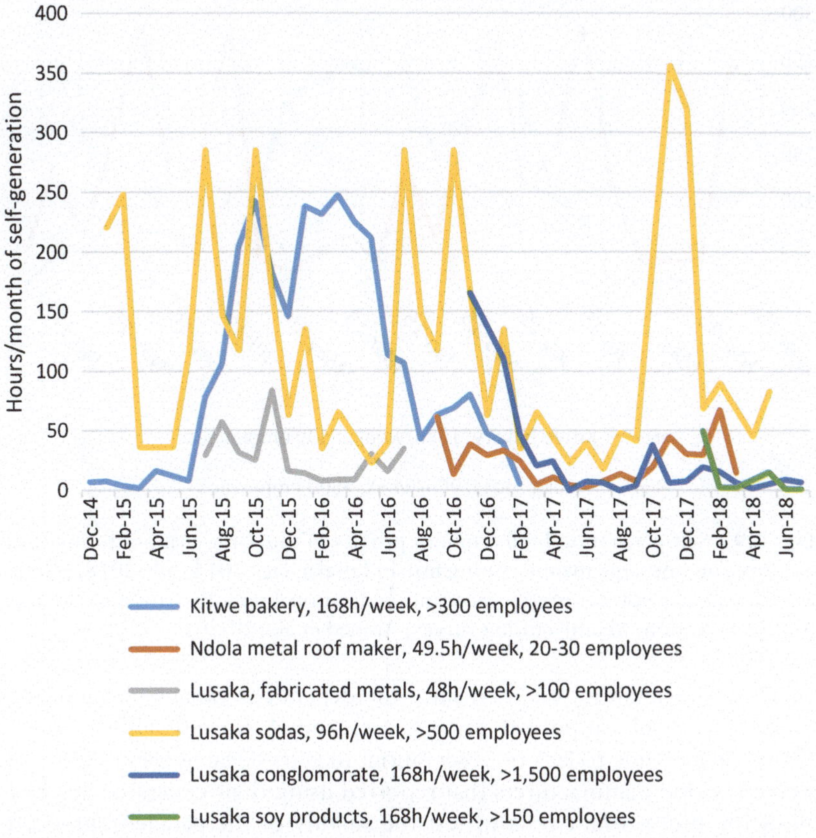

Fig. 4.8 The self-generation hours/month profiles of six large manufacturing firms, Dec 2014-July 2018. (Note: Empty bars do not necessarily mean no generation hours that month. Source: Author's creation, Manufacturing survey; Ahmed et al. (2019))

4.5.1.3 Econometric Observations

Two different types of regression model were run to investigate the predictors of scale of investment in self-generation and extent of use of self-generation, to see how they differed. An ordinary least squares regression model was used to predict the kVA of installed generator capacity that firms invested in by various characteristics, since the dependent variable of kVA installed capacity was continuous. Ordinal logistic regression analyses were run to identify which firm-level variables impacted the extent to

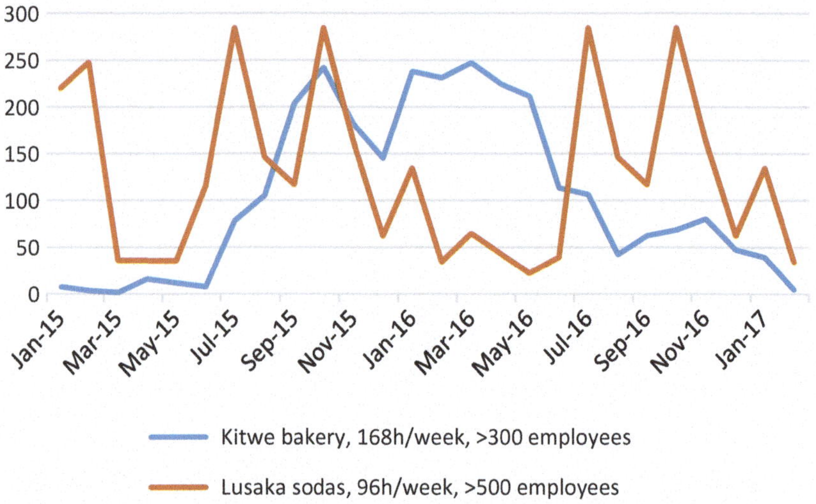

Fig. 4.9 Self-generation hours/month profiles of one large manufacturing firm in Kitwe and one large manufacturing firm in Lusaka, Dec 2014–July 2018. (Note: Empty bars do not necessarily mean no generation hours that month. Source: Author's creation, Manufacturing survey; Ahmed et al. (2019))

which self-generation was used since the dependent variable came in the form of ordinal outputs 0 for no use, and 4 for use all the time. Unfortunately due to lack of observations in pairs of categories (e.g. there were no wood manufacturers that reported using their generators a lot or all of the time), Brant tests to test the validity of the parallel regression assumption could not be run. A generalised ordinal logistic regression model, however, showed that the results held.

Regressions were run against at most seven independent variables at a time, because of the limited number of observations. Below is a table with those variables that were found to have statistically significant associations with the dependant variables, and the models of best fit.

According to the R-squared and pseudo R-squared scores of the two models, the explanatory value of the regression models for predicting self-generation capacity is greater than the models for predicting extent of generator use. This may be a function of richer output data for the former: the installed capacity had a larger range and more diverse output, whereas for the ordinal logistic model, it was just five different outputs, ranging from 0 to 4 (Table 4.5).

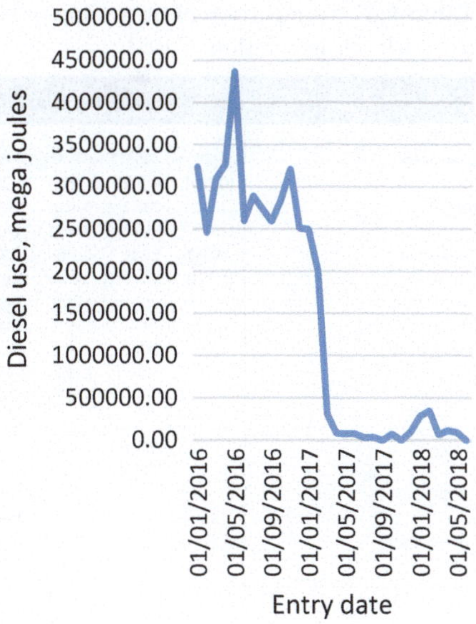

Fig. 4.10 Diesel use by a large beverages company in Lusaka, Jan 2016–Jun 2018. (Source: Author's creation, Manufacturing survey; Ahmed et al. (2019))

The common theme between models 1–5 with installed capacity as the dependent variable is that number of employees, a measure of firm size, was a significant predictor at the 1% level.

Manufacturing production hours is a significant predictor at the 10% level in model 1, where it is one of three independent variables in the regression model. When more variables are added, such as identifiers for whether firms belonged to the basic metals and wood subsectors in model 2 and extent to which backup diesel generation is used in model 5, it loses significance at the 10% level to 10.9% in model 2 and 12.3% in model 5. In models 3 and 4 we lose more than half the number of observations by adding the variable informing whether the firm is Zambian owned. The sample becomes extremely small, and other than number of employees (for which the coefficient changes significantly from models 1, 2 and 5), nothing is shown as significant.

Despite their lower R-squared values, models 6–9 yield more predictors of the extent to which firms self-generate.

In model 6 we have the greatest number of observations, and we see the greatest number of statistically significant predictors: number of employees, whether a firm exports, whether a firm belongs to the basic metals subsector, and whether a firm belongs to the wood subsector

Table 4.5 Comparison of ownership of backup generation capacity versus extent to which self-generation is used

	(1)	(2)	(3)	(4)	(5)	(6)	(7)	(8)	(9)
	Installed generator capacity, kVA					Extent to which self-generate, 0-4			
employees	3.617***	3.639***	6.045***	5.795***	3.432***	0.00288**	0.00682	0.00795**	-0.00141
	(0.509)	(0.517)	(1.295)	(1.282)	(0.509)	(0.00144)	(0.00420)	(0.00402)	(0.00182)
exports	229.2	238.1	438.8	318.8	150.1	0.869**	0.807		0.627
	(157.9)	(159.1)	(431.3)	(430.9)	(158.3)	(0.389)	(0.924)		(0.411)
hoursweek	3.036*	2.891	0.369	-0.375	2.702	0.00199	0.00218	0.00392	-0.000140
	(1.751)	(1.786)	(3.658)	(3.625)	(1.740)	(0.00424)	(0.00798)	(0.00748)	(0.00460)
notzambian			417.9	243.5			1.646**	1.769***	
			(309.4)	(324.8)			(0.692)	(0.679)	
basicmetals		-56.15	-808.4	-532.4	105.0	-2.047**	-6.08*	-6.229*	-2.633**
		(328.6)	(640.3)	(654.5)	(325.4)	(0.818)	(3.224)	(3.218)	(1.340)
o.wood			-	-			1.646**		
							(0.692)		
wood		-353.6			-142.9	-1.809*			-1.484
		(394.5)			(391.9)	(0.932)			(0.941)
selfgen				164.9	117.7***				
				(108.3)	(44.30)				
var(e.selfgen)									
kva									0.00148***
									(0.000447)
Constant	-254.0*	-233.5	-354.7	-415.3	-420.2***				
	(142.3)	(147.6)	(367.0)	(362.7)	(159.9)				
Constant cut4						1.466***			1.324***
						(0.378)			(0.396)
Constant cut1						-0.112	1.375*	1.520*	-0.379
						(0.351)	(0.803)	(0.788)	(0.373)
Constant cut2						0.460	2.169**	2.307***	0.220
						(0.354)	(0.850)	(0.837)	(0.373)
Constant cut3						0.697*	3.975***	4.092***	0.492
						(0.358)	(0.996)	(0.991)	(0.377)
R-squared/Pseudo R-squared	44.8%	45.2%	55.4%	58.2%	48.6%	6.7%	17.1%	17.7%	10.5%
Observations	114	114	42	42	114	119	43	43	114
Regression model	linear	linear	linear	linear	linear	ologit	ologit	ologit	ologit

Source: Manufacturing survey. Author's calculation

Standard errors in parentheses; *** $p < 0.01$; ** $p < 0.05$; * $p < 0.1$

(belonging to both of these subsectors is a negative predictor of extent to which firms use backup diesel generation).

In model 9 we add the variable installed capacity (kVA). This becomes the main explanatory predictor. The effect of number of employees (65% correlation), production hours (38% correlation) and whether a firm exports (29% correlation) become captured within this variable, but whether a firm belongs to the basic metals or wood subsectors do not (1% and –5% correlation).

Table 4.6 Further ordinal logistic regressions involving the basic metals and wood subsectors

	(1)	(2)
Consequences ?		unplanned outages
Coping mechanisms ?	delay	loss of clients
basicmetals	1.898**	1.112*
	(0.888)	(0.656)
wood	-0.103	0.670
	(0.927)	(0.870)
selfgenrecat	-0.999***	-0.513**
	(0.230)	(0.205)
reschrecat	0.553**	0.274
	(0.225)	(0.206)
Constant	2.673***	1.204***
	(0.274)	(0.233)
Constant cut1	-1.934***	-0.430
	(0.447)	(0.371)
Constant cut2	-1.375***	0.551
	(0.436)	(0.368)
Constant cut3	-0.788*	1.603***
	(0.424)	(0.400)
Constant cut4	-0.196	2.428***
	(0.409)	(0.468)
Pseudo R-squared	12.3%	5.1%
Observations	120	119
regression run	ologit	ologit

Standard errors in parentheses

*** p<0.01, ** p<0.05, * p<0.1

Source: Manufacturing survey. Author's calculation

The basic metals subsector was statistically significantly and positively associated with delays in production (model 1 in Table 4.6) and also with a loss of clients when unplanned outages occurred (model 2). The wood subsector was not a statistically significant predictor for either. Brant tests could not be conducted for these ordinal logistic regressions because there were no observations of basic metals firms that experienced no delays in production and there were zero observations of wood across several categories of loss of clients. Generalised logistic regressions however upheld the findings (self-generation was significant only for the difference between small and major delays for basic metals).

4.5.1.4 *Discussion on the Predictors of Coping Mechanisms*

The number of hours a large Zambian manufacturing firm operates is a positive predictor of whether it will keep larger inventories and use capacitors, voltage regulators or power factor correction units. For the former, whether a firm reschedules its workers is also a statistically significant predictor. For the latter, as was discussed already above, damage to inventory was also a predictor.

Timing of acquisition of the use of a self-generator was 92% correlated with when firms reported to have experienced their worst power outages. Tellingly, age—a proxy for a firm's level of experience and business acumen, and one that Steinbuks and Foster found to be a statistically significant predictor of backup generation ownership (Steinbuks & Foster, 2010)—was not a statistically significant predictor of backup generation ownership. This suggests that many large manufacturing Zambian firms did not perceive power outages to be a threat to their operations prior to the outages of 2015 and 2016—indeed this is borne out by the increase in ownership of generators from 38% of firms owning a generator observed from the World Bank's 2002 to 2006 Enterprise Surveys (Steinbuks & Foster, 2010, Table 1) to 73% in 2018 from this work' primary data collection. For the period 2002 to 2006, Steinbuks and Foster had calculated a negative average cost-benefit figure for owning a generator given the infrequency of power outages and the high capital cost of buying one (Steinbuks & Foster, 2010) But during the power crisis of 2015 and 2016, both old and new firms saw the need for backup generators. Timing of crisis took precedent over age of firm.

Like Steinbuks and Foster (2010), the survey found firm size, export orientation and whether a firm was owned by Zambian or not to be statistically significant predictors of generator ownership. My survey analysis in Table 4.5 shows firms with more employees had statistically significantly more installed capacity. It does not show statistical significance for export facing firms or Zambian ownership in this regard, but where it goes beyond Steinbuks and Foster is that it examines predictors of the extent to which firms report self-generation. Firms that exported were likely to report statistically significantly more self-generation use. Firms that were not owned by Zambians, or that were in the basic metals or wood subsectors were likely to report statistically significantly less self-generation use.

Regarding subsectors, the study had different findings than Steinbuks and Foster (2010), perhaps partly due to the lower number of observations, but also more nuanced findings because it went a step beyond.

Like Steinbuks and Foster, the study found that food and beverages were a statistically significant positive predictor of installed backup generation capacity. A former production manager at a milk factory gave a qualitative explanation for why self-generation was important for milk production, which falls within the food and beverages subsector. He said (personal communication, May 25, 2018; Ahmed, 2021) that even half a second of power outage would result in the reset of the manufacturing process for eight hours because of the vulnerable sterilisation process. To prevent this from happening, the firm used ZESCO to charge the batteries for their uninterrupted power supply machines, which they used all of the time. Their machines were not run on ZESCO power directly. This accorded with what Pasha et al. (1989) found: that there is a major variation between the type of industry and the cost of an outage: continuous-process industries are more vulnerable to spoilage and would therefore do the most to protect against losses.

Steinbuks and Foster were able to rank in order of magnitude of coefficient for statistically significant subsectors for most likely to own a generator:

1. hotels and restaurants,
2. food and beverages,
3. chemicals and pharmaceutics,
4. non-metallic and plastic materials,
5. metals and machinery, and
6. wood and furniture.

The survey and analysis showed a similar ranking among subsectors for which there were more than five firms, when ranking subsectors by how frequently their firms marked the use of self-generation by 'most' or 'all' of the time:

1. fabricated metals,
2. food and beverages,
3. textiles and garments,
4. chemicals,
5. plastics and rubber,
6. basic metals and
7. wood and wood products (Fig. 4.11).

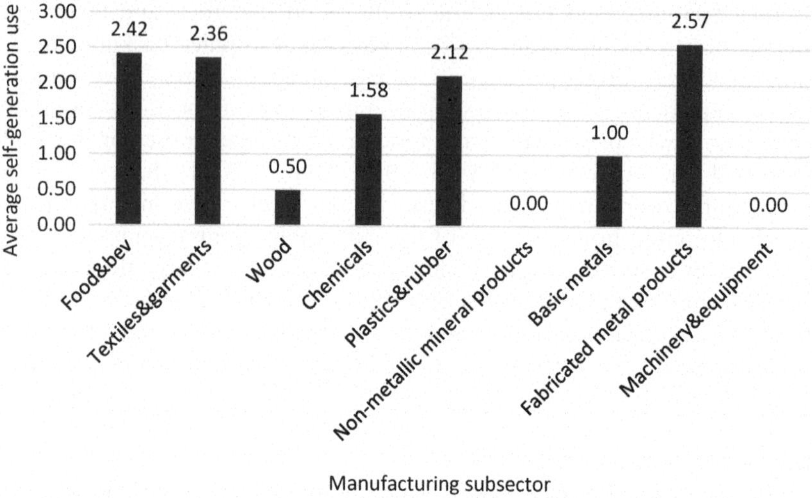

Fig. 4.11 How generation use differs between subsectors. (Source: Manufacturing survey)

Analysis of variance showed the difference in self-generation use between subsectors from the manufacturing survey is statistically different at the 10% level of significance (see Table 4.7).[3]

[3] To carry out this analysis, I

1. calculated the mean for all self-generation scores;
2. calculated the mean self-generation for each subsector;
3. calculated for each value within each subsector the sum of squares difference with the mean for its subsector, and aggregated the 'sum of squares within';
4. calculated for each value within each subsector the sum of squares difference between the subsector mean and the overall mean, and aggregated the 'sum of squares between';
5. found there to be 8 degrees of freedom for sum of squares between and 115 degrees of freedom for sum of squares within.;
6. calculated the *F*-statistic by dividing the sum of squares between by 8 degrees of freedom and dividing this by the sum of squares within and by its 115 degrees of freedom;
7. used an *F*-statistic table (UCLA Statistics, 2019) to find the *F*-statistic at the 10% level of statistical relating to 8 degrees of freedom in the numerator and 115 degrees of freedom in the denominator; and
8. compared this 10% statistical significance *F*-statistic with the *F*-statistic that I calculated.

Table 4.7 How differently firms use self-generation by subsector

	Food& bev	Textiles& garments	Wood	Chemicals	Plastics& rubber	Non-metallic mineral products	Basic metals	Fabricated metal products	Machinery& equipment	Mean	SSW	SSB	df
# firms that self-generate levels 3 or	28	6	0	7	9	0	2	9	0				
% subsector that self-generates leve	61%	55%	0%	37%	47%	0%	22%	64%	0%				
Average self-generation for subsect	2.42	2.36	0.50	1.58	2.12	0.00	1.00	2.57	0.00	**2.07**			
# firms surveyed in subsector	46	11	4	19	19	1	9	14	1				
Square of sums within subsector, SS	139	29	1	57	51	0	18	39	0		**333**		
n-1 degrees of freedom	45	10	3	18	18	0	8	13	0				**115**
Square of sums between subsectors	5.85	0.98	9.80	4.50	0.05	4.27	10.22	3.58	4.27			**44**	**8**
H0: subsectors averages are equal													
H1: subsectors are different													
F-statistic = (SSB/8) / SSW/(115)	1.8776												
Test for alpha = 10%													
For 10% and 8, 120 df, F statistic is =1.77483-(1.77483-			1.72										
Therefore H0 is disproved at the 10% level and H1 is correct													

Source: Manufacturing survey. Author's calculation

Like Steinbuks and Foster (2010), food ranked high, basic metals and wood and wood products at the bottom and the other subsectors ranked in the middle. This study found food and beverages to be a statistically significant predictor of generator capacity, and that the basic metals and wood subsectors were statistically significant predictors that firms categorised as such would *not* self-generate to a great extent.

Unlike Steinbuks and Foster's (2010) study, this one was unable to show statistically significant p-values for all subsectors. Whereas Steinbuks and Foster (2010) relied upon the World Bank's 2010 enterprise survey data for 4246 observations across African nations, this study was limited to 114 firms that shared the relevant data (see Table 4.5). In spite of this, this study yields some richer results. Whereas Steinbuks and Foster's observations came from all over Africa in diverse stages of development, this study focused on Zambia and covered a third of Zambia's large manufacturing firms. Moreover, because this study collected kVA of installed generator capacity—a continuous dependent variable—rather than a binary variable for generator ownership, this study's regression analysis was richer, and so whereas Model 4 in Table 4.5 has an R-squared of 58% for installed generator capacity, Steinbuks and Foster's model had an R-squared of 24% (Steinbuks & Foster, 2010, Table 2). The three ordinal categorical variables (0, 1, 2) used for measuring extent of self-generation use in models 6–9 in Table 4.5, though, did not improve on Steinbuks and Foster's R-squared score.

The study builds upon Steinbuks and Foster (2010) by

- using richer data;
- updating our understanding of the impact of outages on Zambia's manufacturing sector post-El Niño when outages lasted eight hours a day, as opposed to when outages occurred only 5% of the time; and
- considering as Oseni and Pollitt did (Oseni & Pollitt, 2015), that there are still outage losses that are unmitigated by ownership of backup generation. By taking a ground-up approach to asking firms how much capacity they invest in and then to what extent they use their generators to make up for power outage losses, this study allows for the possibility that firms with generators do not completely cover the power losses both because they cannot in terms of capacity and because the firms do not employ them to, because of working capital issues.

By illustrating differences between the statistically significant predictors of installed capacity in Models 1–5 of Table 4.5 and the statistically significant predictors of the extent to which generators are used in Models 6–9, Table 4.5 illuminates stories that may have otherwise been missed. On a subsector level, both wood and basic metals were statistically significant negative predictors of the extent to which firms used their generators. While neither wood nor basic metals were statistically significant predictors of installed capacity of backup generation in this study, a firm belonging to the basic metals subsector was not a statistically significant predictor of *greater* installed capacity whereas belonging to the wood subsector was not a statistically significant predictor of *less* installed capacity. These results suggest different reasons, therefore, for reporting that their use of backup generation was less. The wood and woodwork subsector is not energy intensive (Steinbuks & Foster, 2010), whereas the basic metals subsector is, hence the different directional impact on installed backup capacity. The marginal revenues earned from intensive generator use for the wood and wood products subsector are lower than the marginal costs of self-generating, whereas with basic metals, the issue is perhaps a case of insufficient working capital. Further semi-structured interviews would confirm this. This would be a credible explanation, given Steinbuks and Foster's observation that belonging to the metals and machinery subsector, like belonging to the food and beverages subsector, has a statistically significant benefit-cost margin of owning a generator (Steinbuks & Foster,

2010, Table 5c). (The other subsectors do not have statistically significant cost benefit margins of owning a generator.)

Beyond subsectors, the regression analysis of the extent to which firms used backup generation revealed another statistically significant predictor that was not statistically significant for predicting installed capacity of backup generation: whether a firm was based in Kitwe or not had an impact on the extent to which firms reported self-generation. Because only one enumerator surveyed firms in Kitwe, the statistically significant result could have been due to the particular enumerator's influence on the survey. The result therefore required further investigation. Out of the 18 firms which the particular enumerator interviewed in Kitwe, 83% had generators, which is higher than the 65% average for all firms interviewed in Lusaka, Ndola and Kitwe. Of Kitwe firms with generators, 73% had purchased their generators *prior* to 2015. Recall that in the overall survey of firms in Lusaka, Ndola and Kitwe, 53% of firms had purchased their generators in 2015 or 2016. The Kitwe result therefore seems to be statistically significant not because of the way in which the enumerator filled the surveys, but because of how firms in Kitwe were actually using their generators.

When this result was discussed in a public forum (4 September 2019), an economist at the Energy Regulation Board said it accorded with his own research (Sikwanda, 2019). His research indicates that small firms in Kitwe bore a higher cost of power outages than firms in Ndola, Lusaka and Livingstone (Mwila et al., 2017, p. viii), which would explain why firms in Kitwe might have used self-generation more. The self-generation pattern use of the firm in Kitwe shown in Fig. 4.9, compared with a firm in Lusaka could possibly support this assertion by the Energy Regulation Board (Mwila et al., 2017, p. viii).

Given that there is variance in the extent to which firms use their generators—and some of the factors that impact this are shared by the predictors of installed capacity (i.e. number of employees, whether a firm is export-oriented)—this shows that studies that assume that backup generation mitigates completely against losses have overestimated generation's capability to mitigate against losses, underestimated those losses and would have overestimated emissions associated with backup generation.

The same ranking of subsectors for use of self-generation (where there were more than five firms in the sample) held for highest proportion of firms that used capacitors, voltage regulators or surge protectors all the time:

Table 4.8 To a major extent or all the time and whether they use capacitors, voltage regulators and/or surge protectors all of the time

	Food &bev	Textiles& garments	Wood	Chemicals	Plastics& rubber	Non-metallic mineral products	Basic metals	Fabricated metal products	Machinery& equipment	Mean	SS W	SS B	df
# firms that protect against power surges al	30	6	2	12	10	1	4	9	0				
% subsector that use protection all the time	65%	55%	50%	63%	53%	100%	44%	64%	0%				
Average CVS for subsector, 0-4	2.84	2.27	2.25	2.74	2.67	4.00	2.71	2.77	0.00	**2.69**			
# firms surveyed in subsector	46	11	4	19	19	1	9	14	1				
Square of sums within subsector, SSW	130	40	13	60	48	0	21	44	0		**356**		
n-1 degrees of freedom	45	10	3	18	18	0	8	13	0				**115**
Square of sums between subsectors, SSB	1.05	1.93	0.78	0.04	0.01	1.71	0.00	0.08	7.25			**13**	**8**
H0: subsectors averages are equal													
H1: subsectors are different													
F-statistic = (SSB/8) / SSW/(115)	0.518												
Test for alpha = 10%													
For 10% and 8, 120 df, F statistic is	=1.77483-(1.7748		1.72										
Therefore H0 is not disproven at the 10% level													

Source: Manufacturing survey. Author's calculation

1. fabricated metals,
2. food and beverages,
3. textiles and garments,
4. chemicals,
5. plastics and rubber,
6. basic metals and
7. wood and wood products.

Analysis of variance (see Table 4.8), however, showed that the difference in the use of capacitors, voltage regulators and/or surge protectors between subsectors from the manufacturing survey to not be statistically different at the 10% level of significance.[4]

The same ranking did not hold for willingness to pay more for more reliable electricity (see Sect. 4.6.1.1 for results):

1. wood and wood products,
2. food and beverages,
3. basic metals,
4. textiles and garments,
5. plastics and rubber,
6. fabricated metal products and
7. chemicals.

[4] Using the same methodology as for the above variance of analysis.

Subsectors were not statistically significant in predicting whether a firm would be willing to pay more, so perhaps this should not be looked at too deeply.

Firm size in terms of employees was a statistically significant predictor of both installed diesel generation capacity and the extent to which a firm used its generators. This makes sense because larger firms would have the resources for the required capital expenditure, for the working capital to run their generators and because larger firms would benefit from any economies of scale associated with capital costs and generation capacity (Allcott et al., 2015).

Descriptively, of 88 firms with more than 50 employees, 64 (73%) had self-generation capability. Of 45 companies with fewer than 30 employees, 20 (44%) had self-generation capability. Firms with more than 50 employees were therefore 64% more likely to have a generator than firms with fewer than 30 employees. Similarly, firms larger than 50 employees were 40% more likely to use voltage regulators, capacitors and power surge factor units than firms with less than 30 employees, and therefore see less damage to equipment and inventory. These results accord with the World Bank's, which found that firms employing more than 100 employees felt less of an impact in loss of revenue than firms with 20–99 employees (World Bank and International Finance Corporation, 2014, p. 14).

Exports was a statistically significant predictor of both installed self-generation capacity (if a firm exported, it would on average have approximately 600 kVA of additional installed capacity than if it were a firm that did not) as well as extent to which generators were used. This makes sense because exporters' clients often have a larger pool of potential suppliers for commodities and therefore tend to be more demanding and less tolerant of failure.

Descriptively, of 54 firms that exported, 48 (89%) had self-generation capability. Of 87 firms that did not export, 47 (54%) had self-generation capability. Firms that exported were therefore 65% more likely to have a generator than those that did not. This, and size of firm, triangulates with Steinbuks and Foster's (2010) findings. Hours of weekly production were a statistically significant predictor of a firm's installed capacity; for every additional hour of work, a firm would have on average approximately five additional kVA self-generation capacity. While firms working around the clock have more latitude of when they have to operate their generators (which explains why weekly hours did not appear as a statistically significant predictor of intensity of self-generation use), they also tend to be larger and better resourced.

The results that foreign ownership and years of production experience were not statistically significant factors in determining whether a firm owned and used a generator contrasts with Sichone et al.'s finding that foreign ownership and years of existence had a statistically significant impact on Lusaka-based manufacturing firm productivity (2016). Descriptively, of 36 firms that were foreign-owned, 26 (72%) had self-generation capability. Of 27 domestically owned companies, 13 (50%) had self-generation capability.

Of the 66 firms with more than 15 years of manufacturing experience, 43 (65%) had generators. Of the 39 firms with fewer than seven years of experience, 26 (67%) had generators. The difference in years of experience seemed insignificant. Given that more than half of firms purchased their first generators as recently as in 2015 or 2016, the dire necessity for generators perhaps explains why years of experience was not a factor.

4.6 Costing the Damages for Firms

Section 4.2 showed the most common costs of outages. Section 4.3 showed that using self-generation was the fourth most favoured means of mitigating these costs. This section costs the damages by working out marginal costs of outages by

1. asking firms' production managers and accountants to cost the premium on top of the tariff that their firms paid that they would be willing to recommend their firms pay for a reliable supply of electricity, and
2. observing firms' use of self-generation.

4.6.1 *Marginal Cost of Outages for Firms Inferred from Willingness to Pay for More Reliable Energy*

A quarter of respondents at 119 firms said that their firms would be willing to pay a higher electricity tariff for more reliable energy. Of the three-quarters of respondents who said that they were not willing to pay a premium, they fell into broadly four categories:

1. Those that were satisfied with their arrangements:
   (a) Two responses were that they had secured a bilateral agreement with ZESCO to give them secure energy.
   (b) One response was that the firm now received good communication from ZESCO about outages.
   (c) Three responses were that there was no problem with power outages now (during the period in which the interviews were taken, April–August 2018).
2. Those that opposed tariff hikes on the basis that they were too high for the business (12 firms said that the tariff was already too high or would affect profits).
3. Those that opposed tariff hikes because there had already been two tariff increases in 2017 (2 firms).
4. And those that opposed tariff hikes because they did not trust ZESCO to deliver more reliable energy if they paid a higher tariff. More than a quarter of respondents (32/123) said that they either did not receive notification or received inaccurate and therefore useless notifications. (Mwila et al. (2017, p. 45) found that 17% of businesses said that they did not have access to information for load shedding.)

A total of 52 of 89 negative responses can be attributed to these four broad reasons for not wanting to pay a premium.

Of the 30 firm's representatives that said that they would be willing to pay a higher tariff for more reliable electricity supply, 20 quantified the extra they would be willing to pay. The mean extra, when expunging outlier responses that implied higher tariffs than the cost of self-generation, was 0.61 K/kWh, which at the time of the survey was equivalent to 0.04 USD/kWh. Of 21 firms without a generator, 21% were willing to pay more for electricity, versus of 90 firms with a generator, 26% were willing to pay more.

Because not all firms were willing to pay a higher tariff, predictors of whether a firm would be willing to see an increase in tariff were investigated, and amongst those who would be willing to see an increase in tariff, what increase they would be willing to pay.

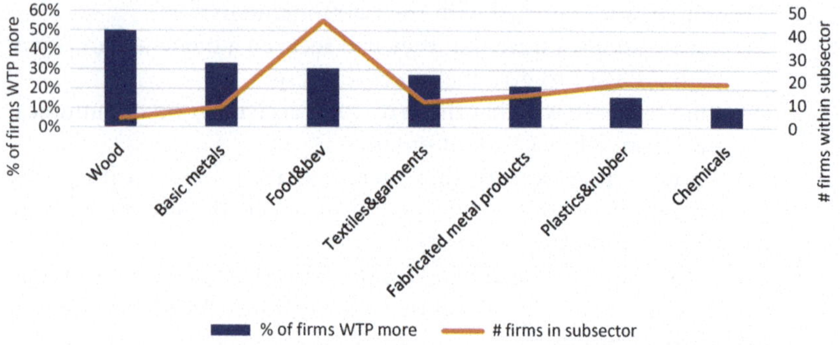

Fig. 4.12 Percentage of respondents willing to pay a higher tariff by manufacturing subsector. (*Source*: Manufacturing survey)

4.6.1.1 Willingness to Pay by Firm Characteristics

Respondents at 25% of firms said that they would be willing to pay a higher tariff for more electricity. Around 50% of wood and wood products firms' respondents said they would be willing to pay more, but the sample of just four firms is too small to reach a strong conclusion. Next most likely was respondents at basic metals firms at 33%, followed by food and beverages at 30% followed by respondents at textiles and garments firms at 27% (Fig. 4.12). These all represented average response rates marginally higher than the overall average. Lower than the average were respondents at fabricated metal products, where 21% said they would be willing to pay more. Respondents at plastics and rubber and chemicals firms were a lot less likely to respond positively to a higher tariff, with 16% and 11% answering in the affirmative (Fig. 4.12).

When whether a firm was willing to pay for more reliable energy was used as a binary dependent variable in logistic regressions, several independent variables were investigated. These variables included firm characteristics (whether a firm was foreign owned, whether it exported, how many employees it had, its revenues, its production hours per week, what subsector it belonged to, its location) and the extent to which firms used various mitigation strategies. The only independent variable that was consistently statistically significant and disproved the null hypothesis was whether a firm exported. More than eight variables could not be modelled at the same time because of the limited number of observations. The food and beverages subsector appeared as statistically significant at the 10% level in a couple of regressions (Table 4.9). The basic metals and wood

Table 4.9 Variables tested for their statistically significant impact on willingness to pay for more reliable grid electricity

VARIABLES	*wtp*
Food/bev	0.975*
	(0.580)
Basic metals	1.312
	(1.006)
Wood	1.779
	(1.124)
Exports	1.442***
	(0.550)
Years	0.0196
	(0.0132)
Hours/week	0.00681
	(0.00560)
kVA	0.000179
	(0.000269)
Employees	–0.00244
	(0.00206)
Constant	–3.109***
	(0.763)
Prob > chi^2	8.2%
Pseudo R^2	11.4%
Observations	111
Model	logit

Source: Manufacturing survey. Author's calculation

Standard errors in parentheses: ***$p < 0.01$; **$p < 0.05$; *$p < 0.1$

subsectors added to the regression model's R-squared and reduced the model's chi-squared probability score, whereas other subsectors had the reverse impacts.

Firm that exports were shown to be willing to pay more for reliable energy at the 1% level of statistical significance. One would expect exporters' clients to be more demanding and less tolerant of failure, especially those with a large pool of potential suppliers.

4.6.1.2 Peak and Off-Peak Usage Indicating Firms' Elasticity of Demand

The enumerators collected ZESCO bills for 48 companies whose respondents said that their firms manufactured for 84 hours of the week or fewer. Their median off-peak energy consumption was 6% and median peak energy consumption was 3%. About 5% of these companies consumed more than 30% of their energy during off-peak hours. None of these companies consumed more than 20% of their energy during peak hours.

The enumerators collected the ZESCO bills for 23 companies whose respondents said that their firms manufactured for more than 84 hours of the week. Their median off-peak energy consumption was 16% and mean peak energy consumption was 9%. Some 9% of these companies consumed more than 30% of their energy during off-peak hours. About 2% of these companies consumed more than 20% of their energy during peak hours.

To test whether the difference in tariffs were insufficient to alter energy consumption behaviour, in April 2019 Field Manager Graham Sianjase called 17 firms that we had earlier interviewed that manufactured 24 hours per day to ask whether there were differentials in the wages that were paid for different times. He got through to three firms and these were the results:

1. A beverages company:
    (a) Normal shift: 8 am–5 pm (wage: normal)
    (b) Changeover shift: 5 pm–7 pm (wage: normal ×2)
    (c) Night shift: 7 pm–7 am (wage: normal ×2)
    (d) Changeover shift: 7 am–8 am (wage: normal ×2)
2. A beverages company:
    (a) Normal shift: 4 am–12 pm (wage: normal)
    (b) Normal shift: 12 pm–10 pm (wage: normal)
    (c) Normal shift: 4 am–10 pm (wage: normal)
3. A steel manufacturer:
    (a) Normal shift: 8 am–5 pm (wage: normal)
    (b) Night shift: 5 pm–7 am (wage: normal + allowance), allowance details not disclosed.
    (c) Changeover shift: 7 am–8 am (wage: normal + allowance), allowance details not disclosed.

Other than knowing that the beverages company with 100% differential in costs in labour for night shifts faces 23% lower electricity tariffs at night than the base tariff, it is not possible to take a view of whether the company is more labour or capital intensive. It has 1500 kVA of self-generation capacity and employs 200 full-time and 115 part-time staff.

4.6.1.3 Energy Consumption Post Tariff Hikes Indicating Firms' Elasticity of Demand

ZESCO increased its tariffs to commercial manufacturing firms twice in 2017—once on 15 May, and once on 1 September. The tariff increases are shown in columns 3–5 of Table 4.10 for firms with Maximum Demand 1, 2, and 3 kVA connections to the ZESCO grid.

ZESCO April and June 2017 bills were shared by eight MD1 firms, four MD2 firms and one MD3 firm and ZESCO August and October 2017 bills were shared by eight MD1 firms, eight MD2 firms and two MD3 firms. While these are not large numbers when split into the three categories, they aggregate to 18 observations for August—October 2017.

Table 4.10 The non-effect of tariff hikes on eight MD1 firms, four MD2 firms and one MD3 firm

1	2	3	4	5	6	7	8	9	10	11
Maximum demand category	Charge for time of day	Tariffs before	Tariffs as of 15 May, 2017	Tariffs as of 1 Sept, 2017	% price increase in May	% price increase May-Sept	average % consumption increase April-June	average % consumption increase Aug-Oct	implied price elasticity, April-June	implied price elasticity, Aug-Oct
MD1-Capacity between 16-300kVA	Energy Charge(K/kWh)	0.2	0.3	0.35	50%	17%	14%	8.5%	0.28	0.51
	Off Peak Energy Charge(K/kWh)	0.15	0.23	0.26	53%	13%	35%	13%	0.66	0.98
	Peak Energy Charge(K/kWh)	0.25	0.38	0.44	52%	16%	49%	12%	0.95	0.75
MD2-Capacity between 301-2,000kVA	Energy Charge(K/kWh)	0.17	0.26	0.3	53%	15%	-1%	11%	- 0.02	0.71
	Off Peak Energy Charge(K/kWh)	0.13	0.2	0.23	54%	15%	16%	15%	0.29	1.02
	Peak Energy Charge(K/kWh)	0.21	0.32	0.37	52%	16%	6%	16%	0.12	1.03
MD3-Capacity between 2,001-7,500kVA	Energy Charge(K/kWh)	0.14	0.21	0.25	50%	19%	1%	1.4%	0.01	0.07
	Off Peak Energy Charge(K/kWh)	0.1	0.15	0.18	50%	20%	18%	12.9%	0.36	0.65
	Peak Energy Charge(K/kWh)	0.17	0.26	0.3	53%	15%	42%	23.6%	0.79	1.53

Source: Author's calculation Manufacturing survey; Ahmed et al. (2019)

Except for a decrease in energy consumption observed from April to June 2017 for energy charged at the standard rate to MD2 firms, an increase in energy consumption is observed for all other types of firm by grid connection size, period and tariff rate. Demand elasticity would be expected to be negative, but this is the case only once in Table 4.10 in columns 10 and 11.

This shows that ZESCO is not charging tariffs high enough to the clients that shared their ZESCO bills tariffs even at the September rate which are impacting their energy consumption. Indeed, energy consumption can be seen to be increasing even more at the peak rate than for the standard rate, especially for the higher Maximum Demand firms, presumably because they are already near their full capacity utilisation rates during standard tariff hours.

These observations and interpretations are matched and supported by Figs. 4.13, 4.14, 4.15, and 4.16, which show the trends in energy consumption for a total of 34 firms that shared their ZESCO bills for the period January 2017 to January 2018.

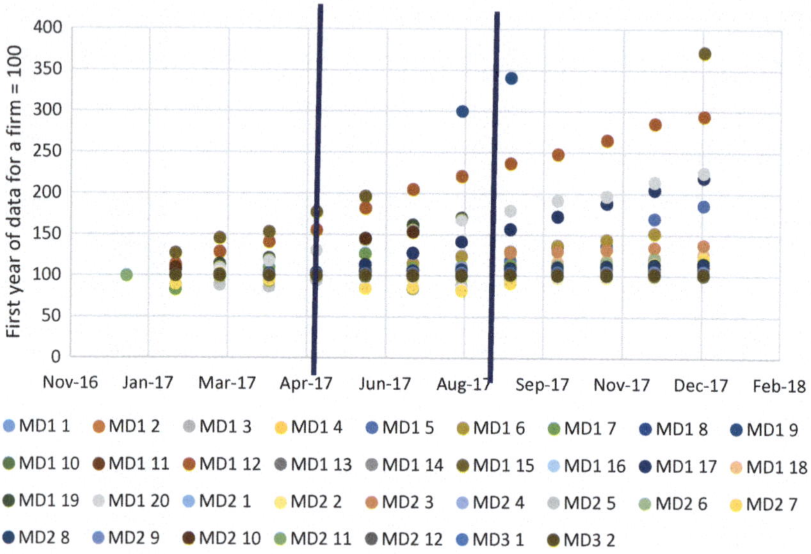

Fig. 4.13 Standard energy consumption for by 34 MD1, MD2 and M3 firms before and after the tariff hikes of 15 May 2017 and 1 September 2017. (Source: Manufacturing survey)

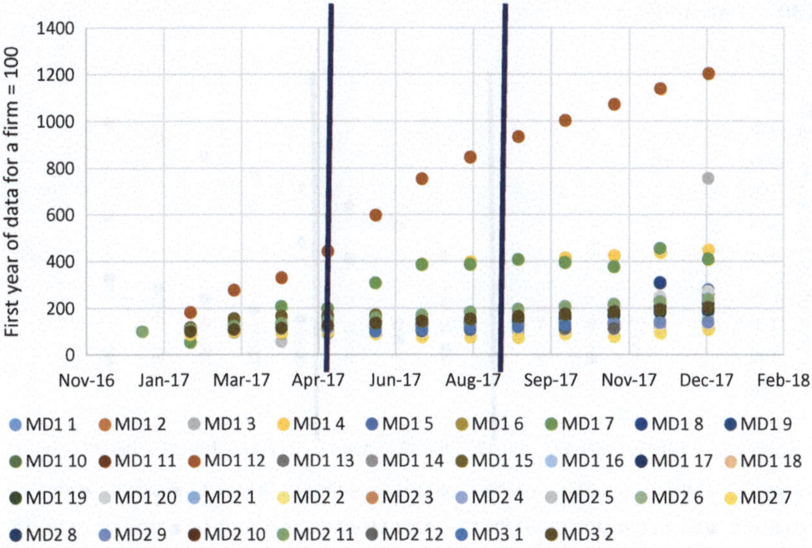

Fig. 4.14 Off-peak energy consumption for by 34 MD1, MD2 and M3 firms before and after the tariff hikes of 15 May 2017 and 1 September 2017. (Source: Manufacturing survey)

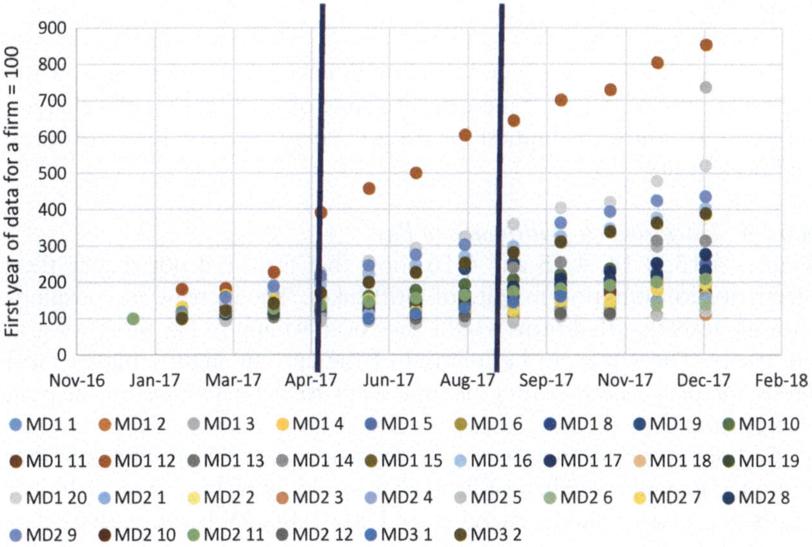

Fig. 4.15 Peak energy consumption for by 34 MD1, MD2 and M3 firms before and after the tariff hikes of 15 May 2017 and 1 September 2017. (Source: Manufacturing survey)

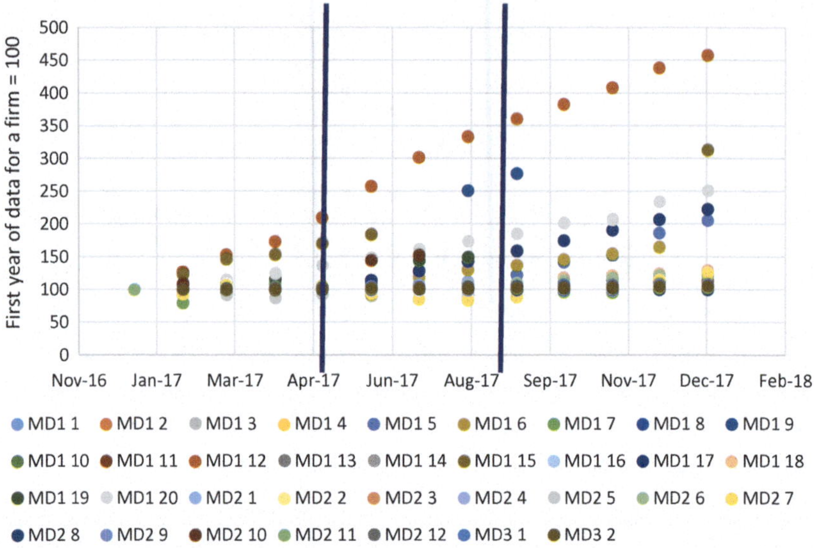

Fig. 4.16 Total energy consumption for by 34 MD1, MD2 and M3 firms before and after the tariff hikes of 15 May 2017 and 1 September 2017. (Source: Manufacturing survey)

Each figure shows a general increase in energy consumption over the period in spite of the tariff increases. In spite of the higher tariffs charged during peak hours, the highest increases in energy use are during off-peak and peak hours.

4.6.1.4 Discussion on Willingness to Pay

Figures 4.13, 4.14, 4.15 and 4.16 show that firms did not reduce their electricity consumption in spite of tariff hikes. The increase in consumption suggests that that firms would have been willing to pay more for grid electricity. The same can be inferred of the current slightly higher tariff levels for peak energy—they do not seem to deter energy use at peak energy hours.

At the time that the surveys were conducted, the average of what respondents said on behalf of firms that would be willing to incur higher tariffs for more reliable energy was USD 0.04/kWh more than they currently did. This is the latest figure offered by research on Zambia, updating the findings of Batizdirai et al., which were made before the most recent tariff hikes (Batidzirai et al., 2018).

USD 0.04/kWh would not likely get ZESCO to its cost recovery levels in order to be able to invest in new installed energy capacity if those cost-recovery levels were USD 0.10/kWh. It needs to be borne in mind that the surveys were conducted in a period of relatively low power outages, but that without an increase in energy supply and Zambia's continued reliance on hydropower, power outages will become a cyclical issue dependent on rainfall and reservoir levels. Firms would likely therefore have expressed higher willingness to pay for more reliable energy during periods of acute power outages. Respondents would also have been inclined to understate their actual willingness to pay in order to protect their firms' profitability. In a separate household survey I carried out online in December 2019 prior to ZESCO's tariff hikes of 1 January 2020 (which were still unlikely to have reached the cost-recovery threshold), 63% of respondents said that they would be willing to pay more for reliable energy—more than double the proportion of respondents on the behalf of manufacturing firms. The average that all respondents said that they would be willing to pay was circa USD 0.095/kWh. The difference between the household and manufacturing surveys were:

1. The timing of the survey: December 2019 was in the midst of even more severe power outages when respondents were experiencing up to 15 hours of official load shedding (ZESCO, 2019), up from eight hours in 2015 and 2016. The timing of the manufacturing survey was April–August 2018, when some manufacturing respondents thought that the power outages were a thing of the past.
2. The way in which I framed the question: For the household survey, I explained to the cost of self-generating energy from diesel generators, explained that a system-sustainable tariff would allow for more reliability and anchored their responses around a cost-recovery mark (see Box 4.1).

Firms most likely to pay for a premium service are those that are export-oriented. ZESCO would therefore have more latitude to invest in increased dependable installed energy capacity on those firms' behalf.

Box 4.1 How the Household Survey Question Relating to Willingness to Pay Was Phrased

15. Households currently pay K 0.15/kWh for the first 200kWh/month and K 0.89/kWh on energy consumed after the first 200kWh/month.

Zesco is not charging any customer a cost-recovery tariff, which means it does not have the money to supply more energy.

Bearing in mind that it costs about K 4.2/kWh to run a diesel-generator, if it meant more reliable energy, what is the maximum you would pay:

- K 1.2/kWh on all your household energy consumption?
- K 1.5/kWh on all your household energy consumption?
- K 1.7/kWh on all your household energy consumption?
- Other, K/kWh (state how much in comments)

Comments:

In terms of firms refusing to accept higher tariffs, distrust of ZESCO's competence seemed to be the most important reason, stemming largely from its inability or unwillingness to communicate correct information about outages. Receiving correct information was the exception rather than the norm. Given that 92% of respondents were located in either industrial zones or a multi-facility economic zone (MFEZ), this doesn't speak well to ZESCO's ability to support manufacturing firms well.

Manufacturing firms felt they were subsidising the mining sector (interview with the CEO of the Zambia Association of Manufacturers on 28 May 2018). The Director of Economic Regulation at the Energy Regulation Board (interviewed on 6 June 2018) said that the Energy Regulation Board was endeavouring to regulate ZESCO's tariff to the Copperbelt Energy Corporation, which was low, by pushing through legislation enabling it to do so. A director of customer affairs said that ZESCO was charging subsidised tariffs to all customers (personal communication, September 4, 2019; Ahmed, 2021).

The second most popular reason cited by 12 firms for not wanting to pay higher tariffs was that it would affect profitability, but judging by the energy consumption rise in peak-hour energy following the two tariff hikes, they could probably afford a further reduction in profitability.

Discussion on Pattern of ZESCO Energy Consumption

Figures 4.13, 4.14, 4.15 and 4.16 show that generally for a sample of more than 30 large Zambian manufacturing firms, energy consumption has been increasing month by month in spite of tariff increases. The same pattern can be seen for 22 firms in Fig. 4.17 from February 2015 until May 2018.

These patterns go against the pattern shown in Table 1.1 where aggregate manufacturing demand dropped in 2016 and had not recovered to its 2015 level by 2016. This might be explained by survivorship bias, as suggested by Sichone et al. (2016)—thus the aggregate amount consumed decreased while the per capita consumption did not.

The month-to-month growth in energy consumption indicates that firms are not producing at capacity and are risk averse: they are expanding production incrementally once demand has been proven, in part at least because of the negative spill-over effects that power outages have on even

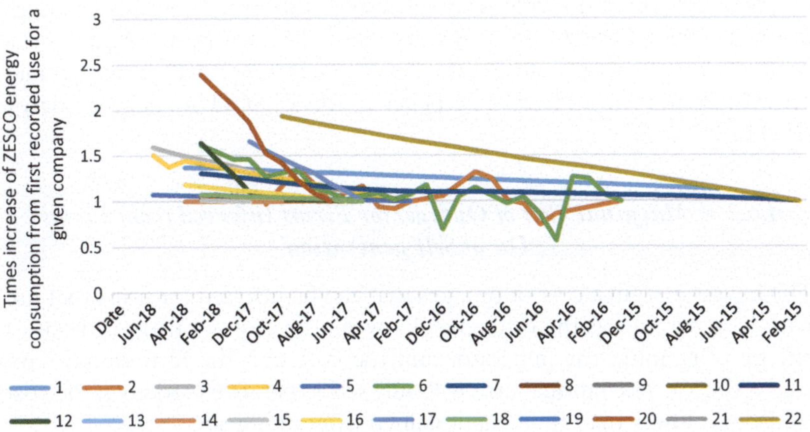

Fig. 4.17 The ZESCO grid electricity consumption patterns of 22 Zambian manufacturing firms. (Source: Author's creation, Manufacturing survey; manufacturing firms' ZESCO bills; Ahmed et al. (2019))

parts of the economy which are not so directly impacted in a costly manner. This is not so dissimilar from what the Ministry of Commerce, Trade and Industry found for firms in 2011–12 (2014, pp. 49–50): firms reported capacity utilisation of 53% in 2006, rising to 66% in 2010, signifying an inefficient use of capital. The Ministry ranked certain subsectors from least efficient in terms of capacity utilisation to most efficient (2014, pp. 50), with textiles (38% capacity utilisation) as least efficient for 2006–2010, and tobacco (88% capacity utilisation) as most efficient.

Beyond the outsized impact of a drop in copper prices and other factors such as exchange rate fluctuations, inflation and access to finance, the 2015/2016 power outages would have slowed growth, and we can see this from the generally gentler slopes of growth of energy consumption in Fig. 4.17 for 2015 and 2016 than the steeper slopes for 2017 and 2018 because (a) from an input perspective, there was less energy available for the firms to use and (b) from an output perspective, demand for their products would have slowed down as other firms were also forced to slow down their expansion.

Another reason for slow expansion in capacity utilisation is what Allcott et al. (2015) predict: rapid economic growth could cause an increase in electricity demand that leads to shortages, which then reduces productivity.

The perpetual and recent growth observed in firms' energy consumption from their ZESCO bills suggests that belief that power capacity is a bottleneck to growth is causing the belief to be a bottleneck to more rapid growth. A way to address that belief is for ZESCO to deliver energy reliably.

4.6.2 *Marginal Cost of Outages for Firms Inferred from Firms' Use of Self-generation*

The lowest marginal cost of outages that can be inferred for firms that use backup generation is the difference between the cost per kilowatt hour of energy of running the generator and the cost that the firm would have paid ZESCO. The rational firm will only self-generate electricity if the cost to do so is lower than the benefit gained from doing so.

It costs approximately USD 0.29/kWh to self-generate electricity (interviews with manufacturing firms and key informants). The variable cost of power outages per kilowatt hour of energy lost therefore has to exceed the differential between this and the cost of power at a given time of day (because there are three tariffs depending on the time of day) for firms to rationally decide to engage in self-generation once they have acquired generators. As of 31 August 2018, when collection of data was finished, the ZMK/USD varied around 13. (It was 18.25 as of 3 August 2020 according to Google.com.) Table 4.11 shows what firms' marginal cost/kWh would have had to exceed for them to self-generate:

Table 4.11 Lowest marginal costs for firms of varying maximum demand capacity to self-generate in 2018

Assumptions:	*13*	*K/USD as of 31 August, 2018*	
	0.29	*USD/kWh variable cost of self-generation*	
Marginal cost of power outages—differential between self-generation and Zesco tariff in 2018			
	% of respondents who fall into this category	*Marginal cost, USD/ kWh*	*% different from standard 16-300kVA*
Maximum demand capacity 16–300 kVA	53%		
Standard charge, 6 am–6 pm		0.263	0%
Off-peak charge, 10 pm–6 am		0.270	3%
Peak charge, 6-10 pm		0.256	-3%
Maximum demand capacity 301–2000 kVA	37%		
Standard charge, 6 am–6 pm		0.267	1%
Off-peak charge, 10 pm–6 am		0.272	4%
Peak charge, 6–10 pm		0.262	-1%
Maximum demand capacity 2001–7500 kVA	9%		

(*continued*)

Table 4.11 (continued)

Assumptions:	*13*	*K/USD as of 31 August, 2018*
	0.29	*USD/kWh variable cost of self-generation*

Marginal cost of power outages—differential between self-generation and Zesco tariff in 2018

	% of respondents who fall into this category	*Marginal cost, USD/ kWh*	*% different from standard 16-300kVA*
Standard charge, 6 am–6 pm		0.271	3%
Off-peak charge, 10 pm–6 am		0.276	5%
Peak charge, 6–10 pm		0.267	1%
Maximum demand capacity 7500+ kVA	0%		
Standard charge, 6 am–6 pm		0.274	4%
Off-peak charge, 10 pm–6 am		0.278	6%
Peak charge, 6–10 pm		0.271	3%

Source: Manufacturing survey. Author's calculation

The table shows that the marginal cost of self-generation reduces during peak charge hours when ZESCO's tariffs are highest because that is when the differential between the cost of self-generation and ZESCO's tariff is lowest. The table also does now show a great deal of variation between off-peak and peak tariffs, nor between the supposedly more generous tariffs to larger electricity consumers. A total of 50% of respondents fell into the category that saw them being charged the highest commercial tariff as the smallest commercial consumers. Thirty-seven percent (37%) of respondents bought energy at slightly lower tariffs for companies with maximum demand capacity of between 301 kVA and 2000 kVA. Some 13% of respondents bought energy at the yet lower tariffs for companies with maximum demand capacity of between 2001 kVA and 7500 kVA.

In Sect. 4.4, it was noted that 95% of firms that had access to a generator used them, meaning most firms with generators incurred a marginal cost of backup generation of at least USD 0.25/kWh. In Fig. 4.4, the most popular years of purchase for generators were identified as 2016 and 2015, when power outages were at their worst. Export-facing firms, firms with more employees, firms with greater production hours and firms in the food and beverages subsector invested in greater capacity. Firms which were based in Kitwe, which had more employees and which were export-facing used their generation capacity more, whereas firms in the basic metals subsector and wood subsector used their generators less, but most likely for different reasons. Given that basic metals firms were likely to invest in more generation capacity but were less likely to use their installed capacity, the marginal cost of outages of circa USD 0.25/kWh might be a huge underestimation for the sector, because it is unable to use the capacity it would want to because of lack of cash flows; firms in the subsector therefore probably suffer more greatly from unmitigated power outages.

4.6.2.1 Discussion

Interviews with key informants and firms indicated that the cost of self-generating electricity is approximately USD 0.29/kWh, not dissimilar to the USD 0.27/kWh figure that Steinbuks and Foster found as the average variable cost of self-generated electricity in Zambia in 2010 (Steinbuks & Foster, 2010). The value of USD 0.59/kWh that Farquharson et al. used (2018), exclusive of capital expenditure, operating and maintenance expenses was likely incorrect.

The minimum marginal cost of outages would be equal to the observed difference between the higher cost of self-generating an extra kilowatt hour of energy and the lost opportunity of paying the lower cost per kilowatt hour of on-grid energy. Previous literature (Bental & Ravid, 1982; Adenikinju, 2003; Steinbuks & Foster, 2010) has got this wrong by including a portion of the sunk capital cost of a generator. Table 4.11 shows that the minimum marginal cost of power outages is north of USD 0.25/kWh, with the largest firms that pay the lowest cost for on-grid energy paying the highest marginal cost of power outages of USD 0.278/kWh for off-peak energy.[5]

[5] As of 31 August 2018, when the data was finished being collected, the ZMK/USD varied around 13.

While it appears to be industrial policy to subsidise Zambia's largest manufacturing firms through the lowest electricity tariffs, the perverse effect is that they could end up paying more on backup generation caused by power outages that could have been avoided had they been charged a tariff that covered the cost of purchasing additional power generation. Consider, for example, the scenario where a large manufacturing firm operating 24 hours per day incurs 8 hours of outages per day, as depicted in Table 4.12. The cost of running backup generation for 100% of its power needs for a third of the day would result in energy costs exceeding those were the firm to pay a cost-recovery tariff to the grid. Beyond the extra energy costs, it would also incur the non-monetised costs highlighted in Fig. 4.2, that is, damage to equipment, reputation, inventory, extra staff costs and loss in clients. In this scenario, the implication is that large manufacturing firms would benefit more from system-sustainable tariffs than lower tariffs.

Besides the issue of costs, there is the issue of affordability. Larger manufacturing firms would be best placed to pay for tariffs that allow ZESCO to deliver reliable electricity. Table 4.5 shows that size in terms of employees is a predictor for both the capacity of installed power and the extent to which generators are used. And yet they pay even less than small manufacturers. Large manufacturers are charged as little as USD 0.01/kWh for off-peak hours while lower maximum demand category commercial customers are charged up to USD 0.03/kWh for peak hour use (ZESCO Ltd., 2017a).

This is not the case in neighbouring Zimbabwe which shares Zambia's Kariba North Dam. This dam accounted for 45% of Zambia's installed power capacity in 2017 and 51% of Zimbabwe's installed capacity in 2016 (CIA, 2016; Energy Regulation Board of Zambia, 2018, p. 63). The

Table 4.12 Hypothetical scenario in which a large manufacturing firm pays more for energy because it is not paying a cost-recovery tariff that would pay for new installed dependable power supply

	Cost, \$/kWh	*Hours*	*Cost (\$)*
Backup generation	0.29	8	2.32
Standard on-grid energy currently MD3	0.016	16	0.26
			2.58
Hypothetical cost-recovery on-grid energy	0.1	24	2.4

Source: Author's calculation

Zimbabwean utility charges manufacturers USD 0.04/kWh for off-peak electricity and USD 0.13/kWh for peak energy (ZETDC, 2014). Zimbabwe is not known for being a paragon of financial prudency, and yet Zambia's public utility makes it look thus by charging lower tariffs based on the marginal operating costs of fully amortised power generation assets, with no accounting for future construction costs. This issue seems to be recognised by ZESCO in its recent application for tariff increases (African Energy, 2019; Phiri, 2019).

Another area for restructuring tariffs is the differential between peak and off-peak charges. Reducing peak hour demand would reduce the pressure on ZESCO to increase its installed power capacity. For example, 84 hours represents operating during the standard energy tariff hours of 06:00–18:00, seven days a week. Firms working more than those hours would necessarily see a higher proportion of their tariffs fall into the off-peak and peak tariff hours. The collected ZESCO bills for these firms show that they consume energy in a 16%:9% ratio for off-peak to peak hour energy. Had they been operating 168 hours a week (the maximum number of hours in a week), they would have consumed at a 33%:17% ratio since 8 hours of the day are off-peak (22:00–06:00) and 4 hours of the day (18:00–22:00) are peak. There is therefore under-consumption of energy during off-peak hours for firms producing more than 84 hours a week but fewer than 168 hours (the maximum number of hours in a week) per week. Figures 4.13, 4.14, 4.15 and 4.16 further highlight the point that peak hour tariffs are doing little to discourage energy during peak hours. Energy use saw its greatest growth in 2017 during both off-peak and peak hours, in spite of two tariff hikes in 2017.[6]

For firms manufacturing 84 hours or fewer, only 5% optimised their energy consumption so that more than 30% of energy consumption fell during off-peak hours. The median off-peak energy consumption was just 6%, suggesting that the differential of K 0.07/kWh (USD 0.005/kWh) between the standard tariff and the peak/off-peak tariffs for 300 kVA and 2000 kVA capacity connections is not enough to incentivise firms to shift their work hours when ZESCO's baseload demand is lowest. By contrast,

[6] Presumably energy consumption growth was greatest during non-standard hours because firms were already manufacturing at capacity during standard hours. To increase energy consumption during standard hours, they would have to install extra capital. It would seem that it is either cheaper or more affordable to increase the capacity utilisation rate during non-standard hours than to install extra capacity for use just during standard tariff hours.

the difference for Zimbabwean manufacturing firms is USD 0.03/kWh between standard and off-peak energy and USD 0.06/kWh between standard and peak-hour energy. Hearing two of three respondents say they pay higher wages for workers working off-peak hours, and hearing one of them state that they pay twice the wages that they do during normal business hours, the difference in cost between off-peak and standard energy would have to off-set this.

4.6.3 *Summary*

There is a large discrepancy between the observed marginal cost to firms of power outages and their stated willingness to pay for more reliable energy. This is to be expected. Running diesel generators does not represent business-as-usual, and indeed the fact that firm size is correlated with generation-use suggests that the costs of self-generation are prohibitive to smaller firms. The discussion above also explored how the cost of backup generation inhibits the extent to which the basic metals subsector use electricity.

A cost-recovery tariff that allows ZESCO to build extra power generation capacity would be better industrial policy than low tariffs that prevent ZESCO from delivering its customers reliable electricity delivery.

To reduce the peak demand it has to meet, ZESCO should charge a much higher differential in tariffs at times of peak demand, so that they alter firm behaviour.

4.7 Special Energy Provision for Important Clients

ZESCO's economists (ZESCO economists, personal communication, November 10, 2017; Ahmed, 2021) said that clients whose power needs exceeded 10 MW or were anticipated to exceed 10 MW, or whose smooth supply of power was seen as a matter of national interest, would be awarded power purchase agreements and prioritised for reliable energy. Enumerators encountered one such company in the light industrial area of Lusaka in the course of our surveys, as well as another company that reported being refused for this service, and instead advised to invest in a transformer. By contrast, contrary to the expectation that multi-facility economic zone (MFEZ) clients would have reliable energy, the one manufacturing firm with which enumerators managed to secure an interview that was located in the south Lusaka Multi-Facility Economic Zone reported four hours of self-generated energy in the week prior to the interview by an enumerator with their assistant supervisor of production.

4.7.1 *Willingness to Sell Energy Back to the Grid*

While prospecting, the Field Manager discovered a cement manufacturer in the Copperbelt that used entirely self-generated energy from its own purpose-built power plant. The firm expressed its interest in selling excess generated energy back to the grid. A total of 29% of 82 respondents said that their firm would be willing to sell energy back to the grid.

4.7.1.1 Discussion

If 29% respondents said that their firm would be willing to sell energy back to the grid, a mechanism should exist for them to do so, especially when ZESCO is forced into enact load shedding.

The team of enumerators did not witness the use of self-generation of energy using solar panels, biomass (although this has been reported in the introduction from the annual report of one listed agro-processing firm) or industrial waste. This represents an opportunity.

A conversation I had with an engineering, procurement and construction contractor and developer of energy solutions for manufacturing firms (2 November 2018; Ahmed, 2021) revealed that they construct 5–40 MW independent power projects that are fuelled by industrial waste such as paper mill pulp (biomass) or such as steel mill blast furnace gas, or excess wood into biomass, or waste into fuel. They are interested in opportunities in Zambia. They said they are one of about ten players operating in a niche space.

An environmental benefit of using biomass, sawdust, tree stumps, papermill sludge, laminated wood residues, deinking sludge, grain, blast furnace gas, and coke oven gas as fuel is that rather than end up landfills, make their way into waterways or be released directly into the atmosphere, these waste products can be used productively to generate energy for the companies that produce them (Martalla, personal communication, November 2, 2018; Ahmed, 2021).

This chapter assessed the results of the manufacturing survey and discussed its contribution to knowledge. Given its industry orientation, the next chapter will investigate why ZESCO had not prepared to prevent the outages that caused such disruptions to the manufacturing sector.

References

Adenikinju, A. F. (2003). Electric infrastructure failures in Nigeria: A survey-based analysis of the costs and adjustment responses. *Energy Policy, 31*(14), 1519–1530. https://doi.org/10.1016/S0301-4215(02)00208-2

African Energy. (2019, March). Zambia: Zesco seeks big tariff increase. *African Energy.*

Ahmed, I., Baddeley, M., Coffman, D., Meikle, J., Oseni, M., & Sianjase, G. (2019). *The cost of power outages to Zambia's manufacturing sector.* https://www.theigc.org/wp-content/uploads/2019/08/Ahmed-et-al-2019-policy-brief-41408.pdf. https://doi.org/F-41408-ZMB-1

Ahmed, I., Parikh, P., Sianjase, G., & Coffman, D. (2020, December 4). The impact decades long dependence on hydropower in El Niño impact-prone Zambia is having on carbon emissions through diesel backup generation. *Environmental Research Letters, 15*(12) https://doi.org/10.1088/1748-9326/abb6a1

Ahmed, I. (2021). *The political economy of energy mix in hydropower dependent developing nations–a case study of Zambia.* University College London. Available at: https://discovery.ucl.ac.uk/id/eprint/10121236/.

Allcott, H., Collard-Wexler, A., & O'Connell, S. D. (2015). *How do electricity shortages affect industry? Evidence from India, NBER Working Paper Series* 19977, Cambridge, MA. https://doi.org/10.1257/aer.20140389

Batidzirai, B., Moyo, A., & Kapembwa, M. (2018). *Willingness to pay for improved electricity supply reliability in Zambia—A survey of urban enterprises in Lusaka and Kitwe.* IGC.

Bental, B., & Ravid, A. S. (1982). A simple method for evaluating the marginal cost of unsupplied electricity. *The Bell Journal of Economics, 13*(1), 249–253. http://www.jstor.org/stable/3003447

CIA. (2016). *Zimbabwe, The World Factbook.*

Diboma, B. S., & Tamo Tatietse, T. (2013). Power interruption costs to industries in Cameroon. *Energy Policy, 62*, 582–592. https://doi.org/10.1016/j.enpol.2013.07.014

Energy Regulation Board of Zambia. (2018). *Energy Sector Report 2017.* http://www.erb.org.zm/downloads/reports/esr2017.pdf

Farquharson, D., Jaramillo, P., and Samaras, C. (2018, October). Sustainability implications of electricity outages in sub-Saharan Africa', Nature Sustainability, 1, 589–597. https://doi.org/10.1038/s41893-018-0151-8 Sustainability.

Government of Zambia. (2008). *The customs and excise (electrical machinery and equipment) (suspension) regulations.*

Ministry of Commerce Trade and Industry. (2014). *Manufacturing Sector Study Report: 2011–2012.*

Mwila, A., Sinyenga, G., Buumba, S., Muyangwa, R., Mukelabai, N., Sikwanda, C., Chimbaka, B., Banda, G., Nkowani, C. and Bwalya, B. (2017). *Impact of load shedding on small scale enterprises.* [online] Lusaka: Energy Regulation Board. Available at: [Accessed 22 June 2021].

Oseni, M. O., & Pollitt, M. G. (2015). A firm-level analysis of outage loss differentials and self-generation: Evidence from African business enterprises. *Energy Economics, 52*, 277–286. https://doi.org/10.1016/j.eneco.2015.11.008

Pasha, H. A., Ghaus, A., & Malik, S. (1989). The economic cost of power outages in the industrial sector of Pakistan. *Energy Economics, 11*(4), 301–318. https://doi.org/10.1016/0140-9883(89)90046-7

Phiri, C. (2019, March). No tariff hike yet, it's only a proposal, says minister. *Zambia Reports.*

Sichone, Y. M., et al. (2016). Electricity load shedding. An econometric analysis of the productivity of firms in the manufacturing sector in Lusaka. *International Journal of Commerce and Management Research, 2*(12), 151–157.

Sikwanda, C. (2019). *Nurturing the manufacturing sector: Ensuring strong roots for Zambian growth.* International Growth Centre.

Steinbuks, J., & Foster, V. (2010). When do firms generate? Evidence on in-house electricity supply in Africa. *Energy Economics, 32*(3), 505–514. https://doi.org/10.1016/j.eneco.2009.10.012

Savin, I. Marson, M., & Sutormina, M. (2020). How different aid flows affect different trade flows: Evidence from Africa and its largest donors. *Structural Change and Economic Dynamics, Elsevier, 55*(C), 119–136.

UCLA Statistics. (2019). *F Distribution tables, statistical online computational resource.* Retrieved April 14, 2020, from http://www.socr.ucla.edu/Applets.dir/F_Table.html

World Bank. (2018). Zambia precipitation 1991–2015. Climate Change Knowledge Portal, World Bank Group. http://sdwebx.worldbank.org/climateportal/index.cfm?page=downscaled_data_download&menu=historical

World Bank and International Finance Corporation. (2014). *Zambia country profile 2013.* World Bank.

Zambia Meteorological Department. (2018). *Raw precipitation data of 39 sites in Zambia, 1991–2018.*

Zesco. (2019). *Countrywide load-shedding schedule October 2019 revised.* http://www.zesco.co.zm/customerCare/loadSheddingSchedule

Zesco Ltd. (2017a). *Current Zesco tariffs, Zesco website.* Retrieved November 2018, from http://www.zesco.co.zm/customerCare/tariffs

Zesco Ltd. (2017b). *Zesco's planned additional energy generation.* Zesco Ltd..

ZETDC. (2014). *Electricity tariffs, ZETDC website.* Zimbabwe Electricity Transmission & Distribution Company.

CHAPTER 5

Why Zambia's System of Energy Provision Did Not Prevent the Power Outages of 2015 and 2016

In Chap. 2 we saw the nexus between industrialisation and economic growth. We were introduced to Zambia's system of energy provision, saw that the World Bank was a significant financier of Zambia's power generation assets in use in 2015 and saw that mineral extraction, beneficiation and industrialisation motivated the World Bank's funding of Zambia's hydropower generation assets, besides its other political interests. In Chaps. 3 and 5 we saw what power outages mean for manufacturing and what they have meant for Zambian manufacturing firms experiencing outages of up to eight hours a day following El Niño-induced droughts. In this chapter we will explain why Zambia's system of energy provision did not prevent the outages of 2015 and 2016, or, indeed, 2019, which saw outages of up to 15 hours (ZESCO, 2019).

Before anything else, we will note the rising incidence of transmission and distribution losses which the literature seems to have neglected. It appears as though these losses surpassed the difference in energy consumption between that in 2014 and the years of the outages in 2015 and 2016. These losses are the first issue to which Zambia's system of energy provision should attend, together with looking at releasing inefficiencies in existing infrastructure through maintenance prior to looking at adding and diversifying power generation assets.

Having assessed the latent additional power generation that can be released without constructing new power generation assets, we will then review the system of provision's demand projections to appreciate to what

I. Ahmed, *The Political Economy of Hydropower Dependant Nations*,
https://doi.org/10.1007/978-3-030-71266-2_5

extent additional power generation is required. Next, we will investigate how climate change and droughts are expected to impact hydropower assets, the Zambian system of provision's favoured mode of generating energy. After that we will consider alternatives sources of energy before finally attempting to explain why Zambia's system of provision did not diversify from hydropower in the Zambezi River Basin in a timely manner. In doing so, we will rely heavily on the historical institutional path dependency framework.

5.1 Transmission and Distribution Losses and Other Inefficiencies

Besides low rainfall which caused a 40% reduction in energy generated at Kariba North hydropower dam—which accounted for 45% of Zambia's installed power capacity in 2016 (Energy Regulation Board of Zambia, 2017a, p. 57)—path dependence and the other factors discussed above, rising transmission and distribution losses since 2007 played a role in Zambia's power outages.

ZESCO's transmission and distribution losses were recorded as 10% (Energy Regulation Board of Zambia, 2017a, p. 63) according to self-reported data by ZESCO. An Energy Regulation Board official noted that this was too good for the kind of data his agency was able to extract from what they were given by ZESCO. The implied losses of generated energy grew to almost a fifth of energy generated in 2016, calculated by subtracting from the output generated the net exports and domestic consumption and dividing that by the output generated (see Table 5.1)—well above the 12% threshold below which ZESCO is required to maintain distribution losses. This could be the result of both transmission losses and non-technical losses, such as pilfering and unmetred use. ZESCO economists said that these calculations in line 6 of the below table were credible calculations (personal communication, 3 September 2019, ZESCO Ltd, Lusaka; Ahmed, 2021). The transmission and distribution losses in 2015 and 2016 were greater than the difference in domestic energy consumption between 2014 and 2015 and 2014 and 2016 respectively.

The literature points to other inefficiencies that can be released. Before investing in additional power capacity, Beilfuss suggests investing in rehabilitation, refurbishment and renovation to increase hydropower capacity (Beilfuss, 2012, p. 36). Yamba et al. advocate for pumped storage schemes

Table 5.1 Energy lost in transmission and distribution compared with the energy deficit in 2015 and 2016

Zambia's energy crisis in 2015 and 2016—decrease in production, increase in energy losses resulting in increased imports

	2013	*2014*	*2015*	*2016*	*2017*	*% drop in 2016 from 2014*
1. Energy output from all sources, GWh	13,299	14,453	13,440	11,696	15,195	*19%*
a. Kariba North	4507	4999	4316	2964	2689	*41%*
b. Kafue Gorge	7463	6666	6417	5853	7363	*12%*
c. Kariba North Bank Extension		1162	1179	672	599	*42%*
d. Victoria Falls	810	811	785	754	684	*7%*
2. Net exports, GWh	1010	1243	391	1391	310	
a. Imports, GWh	73	13	785	2185	753	*-16,708%*
b. Exports, GWh	1083	1256	1176	794	1063	*37%*
3. Domestic energy consumption, GWh	10,846	12,405	11,450	10,858	12,192	
Annual growth in energy consumption		*14%*	*-8%*	*-5%*	*12%*	
4. Output - (exports + domestic consumption), GWh	1370	792	814	44	1940	
5. Output - (net exports + domestic consumption), GWh	1443	805	1599	2229	2693	
6. Implied unused generated energy, % = 5/1	11%	6%	12%	19%	18%	
7. Lost energy, GWh	1443	805	1599	2229	2693	
8. Difference in energy consumption between 2014 and 2015/2016			955	1547		

Author's calculation

Sources: Energy Regulation Board of Zambia (2015, 2016a, 2017a, 2018)

to accompany select hydropower stations to ensure continuous energy (Yamba et al., 2011) by making the most of available water. So far in Africa no hydropower plant has been coupled with solar-pumped storage (Ncube & Lufumpa, 2017). Plans are underway in Zimbabwe to change this. Swedish-owned Ngonyezi Projects has entered a non-consumptive water use agreement with the Zimbabwe National Water Authority to install a

2000 MWh pumped storage hydro facility and 300 MW floating solar photovoltaic plant so that 30% of the solar PV is supported by storage. The solar panels are anticipated to float on 500 hectares of reservoir surface. Additional advantages associated with having the solar panels float on the reservoir surface is that the water will cool the solar panels, reduce algae growth and reduce evaporation (ESI-Africa, 2019). Rehabilitation, refurbishment, renovation and reusing available water resources with pumping should take priority over plans to build new capacity.

5.2 Zambia's Future Grid Electricity Demand

Zambia's energy demand can be forecast by applying the elasticity between GDP and energy consumption. A log-log regression of Zambian energy consumption (Energy Regulation Board of Zambia, 2016a, 2017a, p. 9, 2018, p. 36, 2019a, p. 39; World Bank, 2020b) against Zambia's real GDP (World Bank, 2020c) and Zambia's population from 1990 to 2018 (World Bank, 2020c) shows that GDP is a statistically significant predictor (see Table 5.2). A 1% increase in GDP is associated with a 0.52% increase in energy consumption.

Table 5.2 Log-log regression of the relationship between energy consumption, GDP and population on Zambia's energy consumption

Variables	*1990–2018* *ln energy*
ln GDP	0.516***
	(0.182)
ln population	0.0260
	(0.356)
Constant	-4.305**
	(1.617)
Observations	29
Prob. > *F*	0.00%
R-squared	89.1%

Author's calculation

Standard errors in parentheses

***$p < 0.01$, **$p < 0.05$, *$p < 0.1$

Data source: Energy Regulation Board of Zambia (2016a, 2017a, p. 9, 2018, p. 36, 2019a, p. 39), World Bank (2020b)

Forecasting is coarse because of lack of historic data on energy consumption by income bracket. The World Bank's Multi-Tier Framework collection of data by five tiers of electricity consumption will solve this problem for future data analysis. Beyond the World Bank, government agencies provided electricity access data. It was apparently 22% in 2008 (Ministry of Energy and Water Development, 2008) for all forms of access and 32.9% in 2018 (ZICTA and CSO, 2018). It is difficult to pinpoint when off-grid solar home systems became prevalent in Zambia, but it is probably safe to assume that most electricity accessed in Zambia in 2008 would have been via the grid.[1] Going forward, ZESCO may want to attempt to aggregate energy consumption by when household consumers first received their grid connections.

Based on extrapolating the IMF's economic forecast, we can coarsely forecast Zambia's energy demand by 2030. The IMF's World Economic Outlook used statistics from Zambia's National Statistics Office, which provided last actual[2] data from 2017 (IMF, 2019). The pre-COVID-19 and pre-power outages of 2019 and 2020 forecast the IMF published was that Zambia's real GDP in 2024 would be 115% of its real GDP in 2017, a compounded annual growth rate (CAGR) of 2%. On the basis of 13,080 GWh energy consumed in 2018, a sustained 2% CAGR GDP growth from 2018 to 2030 would imply energy demand of 16,600 GWh by 2030 by applying the elasticity in Table 5.2. This is before accounting for greater electricity connectivity (for what is the point in worrying about Zambia's economic growth if it does not result in better living standards for the majority of the country's population?), but also before accounting for greater energy efficiency.

[1] The Ministry of Energy and Water Development's 305 page Power System Development Master Plan for Zambia published in 2011 mentions solar home systems only once on p. 6, and then only in terms of knowledge of demand for them (Ministry of Energy and Water Development, 2011, p. 6). As of 2010, off-grid solar devices were serving only 1 million households worldwide, compared with 73 million households by 2017 (Dalberg Advisors, Lighting Global and World Bank, 2018, p. 28), when 86% of off-grid suppliers were concentrated in East Africa, compared with just one supplier (Fenix International) in Zambia (Dalberg Advisors, Lighting Global and World Bank, 2018, p. 95). At the turn of the century, off-grid solar systems were not yet technologically developed to be cost-effective without governments subsidies in Argentina and Sri Lanka (World Bank, 1999; Gunaratne and LGA Consultants, 2002). By 2012 in India, though, they were proving cheaper than kerosene for household lighting (Buragohain, 2012).

[2] As opposed to estimated.

If we were to assume that those who are not currently connected to the grid are excluded from GDP forecasts, then greater grid connectivity would further add to Zambia's grid electricity demand. Greater grid connectivity can be driven by market forces, by political will or both, which in turn can be driven by international donors or the domestic electorate. Neither market forces nor political forces will seem theoretically to support greater expansion in on-grid connectivity.

In terms of international donor political will, the United Nations Sustainable Development Goal 7 for affordable, reliable, sustainable and modern energy for all by 2030 does not need to be met by on-grid energy; it can be met by off-grid solar products (Lighting Global, 2020). Off-grid solar solutions seem to 'leapfrog' traditional on-grid electrification in the way that mobile phones did land-lines across Sub-Saharan Africa because they overcome the lumpy, indivisible and irreversible barriers to private investment in generation and distribution (Ahmed, 2017). In fact, World Bank institution Lighting Global and Vivid Economics consider Zambia to be the world's eighth most attractive market for pay-as-you-go solar energy products (Lighting Global and Vivid Economics, 2019). It seems that the donor community's comfort with off-grid solutions absolves the Government of Zambia of the responsibility to connect its population to the grid.

In terms of domestic political will, Zambia's last National Energy Policy was written in May 2008 (Ministry of Energy and Water Development, 2008) (emailed to me by an economist at the Energy Regulation Board on 9 September 2019; Ahmed, 2021) and its latest Power System Development Master Plan for Zambia was released in 2011 (Ministry of Energy and Water Development, 2011). As of 2008, 22% of the population had access to electricity (Ministry of Energy and Water Development, 2008), and the objective 'to increase access to electricity' (Ministry of Energy and Water Development, 2011, p. 19) was vague, with no specific, measurable or time-bound target set. This seems to indicate political inertia on the matter of expanding grid electricity access which as of 2018 was just 6% in rural areas (ZICTA and CSO, 2018). Nor do rural areas generally constitute the voter base for the populist Patriotic Front party (Resnick and Thurlow, 2014). In terms of market forces, Zambia's growth trajectory has favoured the elite, as illustrated by a whole swathe of metrics in Sect. 2.4, and the election of the Patriotic Front on a platform against inequality in urban areas.

Yet Zambia did see an increase in access to grid electricity with probably electoral pressure on Michael Sata's administration to deliver. From electricity access to 2.8 million people in 2008,[3] Zambia's access to grid electricity access increased to 5.6 million people by 2018,[4] representing an increase of grid electricity access for at least[5] 2.7 million people over ten years.

If ZESCO were to extend grid electricity to another 2.7 million additional Zambians, or 530,000 households (there are on average 5.1 Zambians per household (United Nations, 2017)) over a period of 12 years by 2030, and each household consumed on average 11 kWh/day,[6] annual grid electricity consumption would increase by 2100 GWh from among people who are not currently grid-connected. Together with the existing base demand and increase in electricity consumption driven by an increase in GDP of 16,600 GWh, this grid expansion would make grid electricity demand 18,700 GWh by 2030, assuming an ability of new users to pay ZESCO's tariffs.

By contrast, the Japan International Cooperation Agency, a donor agency, forecast that by 2025, Zambia's energy demand would be 18,000 GWh and would reach 21,000 GWh by 2030 (Japan International Cooperation Agency (JICA) and JERA Co. Inc., 2017, tables 5.1–9).

In fact, JICA's forecast was based on an erroneous base of 14,370 GWh consumed in 2016—3500 GWh more than what was consumed in Zambia according to the Energy Regulation Board (2017a, table 2–1). This difference in baseline is more than the difference between JICA's forecast and the forecast of 18,700 GWh.

Taking into consideration the excessive losses in transmission and distribution, if ZESCO were to reduce implied energy loss from 18% in 2017 (line 6 in Table 5.2) to 10%, it would regain 1200 GWh of energy.[7] 15,195 GWh output in 2017 from the Energy Regulation Board's counted power generation assets less 10% of transmission and distribution losses is 13,700 GWh. This means Zambia's system of energy provision would

[3] 22% access (Ministry of Energy and Water Development, 2008) × 12.8 million people (World Bank, 2020c)

[4] 33% (ZICTA and CSO, 2018) of 16.9 million people (ZICTA and CSO, 2018)

[5] Since some of those with electricity access in 2008 may have had off-grid electricity access.

[6] I was unable to find a statistic for average Zambian off-grid energy consumption. 10.75 kWh was the average of summer and winter off-grid energy consumption in Uttar Pradesh (Sandwell et al., 2016).

[7] 15,195 GWh in 2017 * (18%–10%) = 1174 GWh

need to generate an additional 5000 GWh energy by 2030 relative to what it generated in 2017.

5.3 Droughts in Zambia: Forecasts, Historical Occurrences and Competition for Water

Having assessed how much additional energy Zambia needs from new power generation assets, we will assess the suitability of hydropower as a continued source.

Whereas political opposition argued that the power crisis was caused by the government's poor water and energy management, the government blamed climate change (Mfula, 2016; Funder et al., 2018).[8] The two explanations are not mutually exclusive, but in fact go hand-in-hand if one extends the meaning of government from the political party which forms the executive to government institutions, and the period over which these institutions are scrutinised from an election cycle to a longer time horizon. We will use the historical institutional path-dependency framework to explain why Zambia's system of energy provision did not respond to the changing realities of climate change.

Forecasts since 2002 have been indicating that as global temperatures rise, the frequency and intensity of drought and low reservoir levels would reduce Zambia's hydropower capacity factors (Harrison and Whittington, 2002; Beilfuss, 2012; Spalding-Fecher, 2012). The power outages of 2015 and 2016 bore this out. In 2016, the average capacity utilisation rate across Zambia's four large hydropower dams fell from 71% in 2014 to 54% in 2016 (see Table 5.3).

Yet instead of heeding the warnings, as well as the Intergovernmental Panel on Climate Change's forecasts (IPCC, 2018) for a 1.5 °C and 2 °C above pre-industrial levels scenarios, ZESCO neither pursued with urgency the diversification from hydropower with the commissioning of the Maamba Collieries coal-fired power plant in the half decade prior to 2015, nor had it changed its plans to not commission additional hydropower after 2015. Table 5.4 shows that as late as November 2017, ZESCO was planning to commission more than 70% of additional installed power capacity to come from hydropower (Zesco Ltd, 2017b).

[8] The Energy and Water Development minister said 'The situation is not only affecting Zambia alone but the entire southern African region due to Climate Change' (Mfula, 2016).

Table 5.3 In 2016, Zambia's large hydropower dams' average capacity utilisation rate was 54%

Average capacity utilisation rates of Zambia's large dams in 2015 and 2016

	Kafue Gorge Dam	*Kariba North Dam*	*Kariba North Ext Dam*	*Victoria Falls Dam*	*Weighted average capacity utilisation rate*
Installed capacity as of 2014, MW	990	720	360	108	
Energy generated in a year at 100% capacity utilisation, MWh = capacity × time	8,672,400	6,307,200	3,153,600	946,080	
Hours/y	8760				
Generated energy, 2014, MWh	6,666,000	4,999,000	1,162,000	811,000	
2014 capacity utilisation, %	77%	79%	37%	86%	71%
Generated energy, 2015, MWh	6,417,000	4,316,000	1,179,000	785,000	
2015 capacity utilisation, %	74%	68%	37%	83%	**67%**
Generated energy, 2016, MWh	5,853,000	2,964,000	672,000	754,000	
2016 capacity utilisation, %	67%	47%	21%	80%	**54%**
Generated energy, 2017, MWh	7,363,000	2,689,000	599,000	684,000	
2017 capacity utilisation, %	85%	43%	19%	72%	59%

Author's calculation

Data source: Energy Regulation Board of Zambia (2015, 2017a)

Additionally, while ZESCO did increase electricity tariffs to end-consumers, it did not do so to a level that would allow it to finance the commissioning of additional assets (Haria & Ahmed, 2020).

The remainder of this chapter will build out the details of this argument. It will show historic rainfall patterns across Zambia, show the historic analysis that was available to ZESCO when it was making its decisions, namely, Maamba, and show the IPCC's projected climate outlook for

Table 5.4 Zesco's power generation expansion plan as of 2017

Zesco's power generation expansion plans, as of 10 November 2017

Power plant	*Developer*	*Energy source*	*Capacity, MW*	*Project status*	*Commissioning year*
Maamba I	Nava Barat	Thermal (coal)	300	Construction	2016
Maamba II	Nava Barat	Thermal (coal)	300	Feasibility	2021
EMCO I	EMCO Energy	Thermal (coal)	300	Feasibility	2022
EMCO II	EMCO Energy	Thermal (coal)	300	Feasibility	2022
Lunsemfwa/Mkushi	LHPC	Hydro	190	Feasibility	2023
Kabompo Gorge	CEC	Hydro	40	Construction	2022
Mulungushi	LHPC	Hydro	50	Feasibility	2018
Lusiwasi Upper	Zesco	Hydro	15	EPC contractor mobilisation	2019
Lusiwasi Lower	Zesco	Hydro	86	EPC contract awarded	2021
Kafue Gorge Lower	Zesco	Hydro	750	Construction in progress	2020
Musonda Falls	Zesco	Hydro	10	Construction in progress	2017
Chishimba Falls	Zesco	Hydro	15	Construction in progress	2018
Batoka Gorge	Zesco	Hydro	1200	Financing	2023
Lusaka	Ngonye	Solar	74	Financing	2019
Kitwe	IPP	Solar	100	Pre-feasibility	2019
Other towns	TBA	Solar	100	Pre-feasibility	2019
Mutinondo	IPP	Hydro	43	Feasibility completed	2021
Luchenene	IPP	Hydro	34	Feasibility completed	2022
Kabwelume Falls	LPA	Hydro	96	Feasibility completed	2023
Kumdabwika Falls	LPA	Hydro	151	Feasibility completed	2023
Luapula-CX-Mombututa	Zesco/SNEL	Hydro	301	Feasibility in progress	2022

(*continued*)

Table 5.4 (continued)

Zesco's power generation expansion plans, as of 10 November 2017

Power plant	*Developer*	*Energy source*	*Capacity, MW*	*Project status*	*Commissioning year*
Luapula-Mambilima-2	Zesco/ SNEL	Hydro	206	Feasibility in progress	2024
Luapula-1-Mambilima	Zesco/ SNEL	Hydro	120	Feasibility in progress	2025
Wind power	Zesco/ MASEN	Wind	150	Pre-feasibility in progress	2022
Solar power	Zesco/ MASEN	Solar	200	Pre-feasibility in progress	2020
GETFIT	IPP	Solar	50	Pre-feasibility in progress	2020
Kalepela	Zesco	Hydro	4	Feasibility in progress	2023
Kapamba	Zesco	Hydro	12	Feasibility in progress	2024
Namundela	Zesco	Hydro	5	Pre-feasibility in progress	2024
Mumburuma	Zesco	Hydro	8	Pre-feasibility in progress	2024
Devil's Gorge	Zesco/ ZRA	Hydro	543	Pre-feasibility planned	2030
Mpata Gorge	Zesco/ ZRA	Hydro	1000	Pre-feasibility planned	2035
Total			**6753**		

Summary of planned projects by energy source by 2035 capacity	**2022 capacity, MW**	**2022 % share**	**2035 capacity, MW**	**2035 % share**
Hydro	1344	42%	4879	72%
Thermal (coal)	1200	37%	1200	18%
Solar	524	16%	524	8%
Wind	150	5%	150	2%
	3218	100%	**6753**	100%

Source: Zesco (2017b)Author's calculation

Zambia. It will introduce the nexus approach for evaluating the use of water for drinking, agriculture and power generation. It will then introduce alternative sources of energy at ZESCO's disposal. Following this context, the remaining chapter will introduce and incrementally build upon the historical institutional path-dependency framework to explain how early decisions have disproportionate consequences for the evolution of later decisions (Coffman, 2013). The remaining chapter will then apply the historical institutional path-dependency framework to explain why ZESCO did not diversify away from hydropower energy as domestic demand was catching up with supply and as the threat of drought was starting to be discussed. It will then explain how the pursuit of supporting Zambia's sectors driving economic and industrial development through subsidies and complacency for decades in not having to recover capital costs for fully depreciated generation assets resulted in ZESCO's inability to directly fund new power generation projects, or credibly pay independent power projects (IPPs) cost-recovery tariffs.

* * *

The IPCC Special Report on the impacts of global warming of 1.5 °C above pre-industrial levels states that there is high confidence that human-induced global warming has resulted in an increase in the frequency and duration of marine heatwaves, and that further, there is medium confidence that it has led to an increase in the frequency and intensity of heavy precipitation events and increased risk of drought at a global scale (IPCC, 2018, p. 177). El Niño events fall into the latter category. Southern Africa is one of the regions associated with increased drought (IPCC, 2018, p. 178). With global temperatures rising, the El Niño effect is predicted to increase in frequency (Wang et al., 2017). Models agree that for southern Africa, temperatures will rise faster at a range of 1.5 °C–2.5 °C as compared to the global warming average of 0.5 °C–1.5 °C (IPCC, 2018, p. 197). At a 1.5 °C temperature rise, 'a robust signal precipitation reduction is found over [...] areas of the Zambezi basin in Zambia' as well as in other parts of southern Africa (IPCC, 2018, p. 197). At a 2 °C temperature rise, the region is projected to face 'robust precipitation decreases of 10–20% and increases in the number of [consecutive dry days]' (IPCC, 2018, p. 197). Further, reductions in streamflow of 5–10% in the Zambezi River basin have been associated with increased evaporation and transpiration events resulting from a rise in temperature, wish issues for hydroelectric power across southern Africa (IPCC, 2018, p. 197).

If IPCC forecasts were not enough to make future reliance on hydropower seem unattractive, even the historical base of evidence shows that

a. the reduction in rainfall in 2015 and 2016 was not unprecedented: 20% of years between 1991 and 2014 had seen less precipitation, and the level of precipitation in 2015 was less than a standard deviation from the mean precipitation for 1991–2014 (see Fig. 5.1). World Bank data on precipitation for an unspecified location in Zambia for 1991–2015 (World Bank, 2018) showed that 2015 was worse than average years for precipitation, but not unforeseeably bad.[9] Compounding the low reservoir levels for 2016 would have been the low levels from the year prior, but again, there was precedence for this (and the low rainfall in 2019—

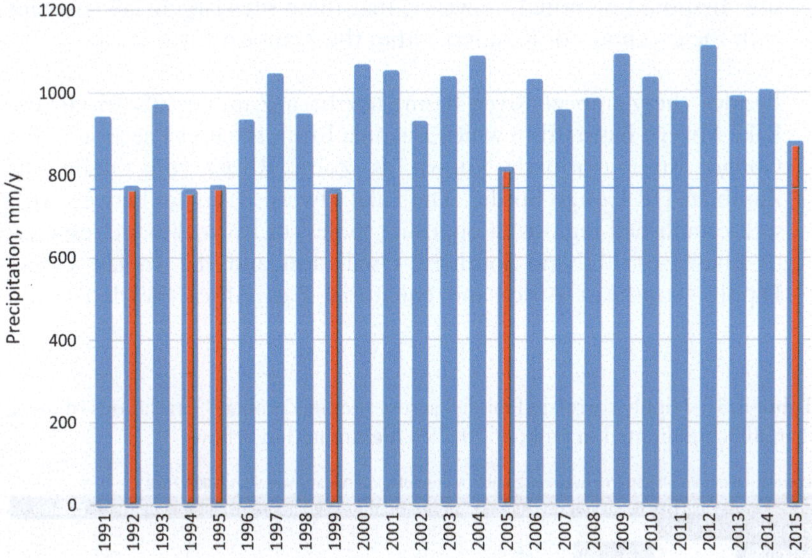

Fig. 5.1 Zambia's precipitation levels for an unspecified location. Almost a quarter of the 25 years 1991–2015 saw rainfall equal to or less than in 2015. (*Source*: World Bank, 2018; graph: own)

[9] As Zambia Meteorological Department data presented later in this work show, precipitation results do not vary greatly across Zambia, and nor does it vary much across the Southern African Power Pool. Zambia lies in the middle of the Zambezi River Basin, which also covers Angola, Botswana, Malawi, Mozambique, Namibia, Tanzania and Zimbabwe.

three times in five years) in the past 25 years: 1994 and 1995 had been two consecutive years when rainfall had been lower than it was in 2015, and counting 1992 made three years of low rainfall in five years. The power outages of 2015 and 2016 were therefore a function of two things happening simultaneously: low-rainfall constrained supply from hydropower in 2015, and demand for energy at its highest level yet.

b. Zambia's geography is largely not resilient to drought: of 39 sites for which the Zambia Meteorological Department has rainfall data, only 9 have data for recent precipitation patterns, the sites for which are indicated on the map below. In general rainfall patterns have at least a 70% correlation (Zambia Meteorological Department, 2018) (see Table 5.5), the only exception being the Kabompo site which saw correlation as low as 65% with the Solwezi site and 69% with the Kasepma site. It should be noted, however, that these sites largely fall (perhaps with the exception of Kasama) within the Zambezi River Basin.

Besides the Zambezi River Basin, northern Zambia falls within the Lake Mweru Basin, from which Luapula River, the upper section of the Congo River (formerly known as Zaire River) runs along the Zambian-DR Congo border. The Lake Mweru Basin lies between two well-established regions of opposing 'poles' of hydroclimatic response to phases of El Niño-Southern Oscillation and the Indian Ocean Dipole—southeast Africa and equatorial East Africa (Waylen et al.,

Table 5.5 Highly correlated rainfall across sites in Zambia. Correlations of more than 80% highlighted in pink; of 70–79% highlighted in yellow

Correlation in rainfall patterns between Zambia Metereological Department sites, 2000-2018

Observations	Zambezi	Kabompo	Kasempa	Solwezi	Lusaka 1	Lusaka 2	Serenje	Mfuwe	Kasama
Zambezi	202	76%	75%	79%	82%	81%	79%	82%	76%
Kabompo	76%	80	69%	65%	73%	71%	74%	76%	76%
Kasempa	75%	69%	88	87%	80%	81%	83%	71%	85%
Solwezi	79%	65%	87%	191	79%	76%	86%	73%	83%
Lusaka 1	82%	73%	80%	79%	197	91%	80%	82%	74%
Lusaka 2	81%	71%	81%	76%	91%	191	79%	79%	71%
Serenje	79%	74%	83%	86%	80%	79%	182	78%	83%
Mfuwe	82%	76%	71%	73%	82%	79%	78%	197	73%
Kasama	76%	76%	85%	83%	74%	71%	83%	73%	197

Author's calculation

Data source: Zambia Meteorological Department (2018). Map made with the assistance of Graham Sianjase

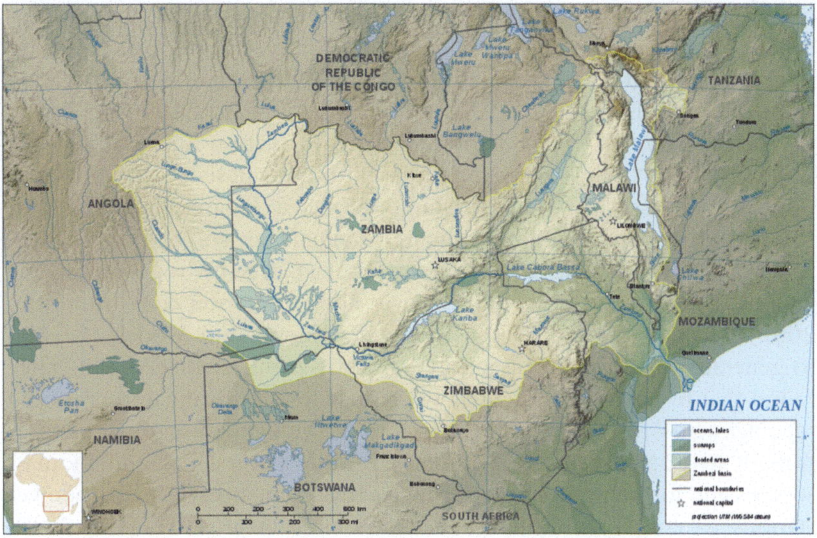

Fig. 5.2 The Zambezi River Basin covers most of Zambia. (*Source*: By Eric Gaba—Wikimedia Commons user, https://commons.wikimedia.org/w/index.php?curid=41919543)

2019, p. 11). Attempting to deduce that El Niño- and the Indian Ocean Dipole-induced opposing effects on annual precipitation and Lake Mweru levels, Waylen et al. noted that the varying numbers and locations of stations used to estimate basin input were a weakness and made their results tenuous (Waylen et al., 2019, p. 10). Whereas the IPCC has consistently forecast reduced precipitation and water availability levels in 2001 and 2018 for the Zambezi River Basin (Fig. 5.2), it has produced contradictory forecasts for the Congo River Basin.

As early as 2001, the literature has been forecasting reducing reductions in streamflow along the Zambezi. In 2001, the IPCC had categorised the Zambezi as the 'worst' of 11 major African river basins in terms of exhibiting potential effects of climate change, with rainfall expected to decrease by 10–20%, increases in temperature expected to result in increased evaporation by 10–25%, resulting in 26–40% reduced average annual streamflow (IPCC, 2001, table 10–1) by 2050 (Beilfuss, 2012, p. 3). In 2002, Harrison and Whittington had sounded the alarm bells for the proposed scheme for a hydropower station at Batoka Gorge's susceptibility to climate change—a scheme that Zambia's system of provision was making progress on preparations for in 2018, and had in fact extended from an envisioned 1200 MW in 2017 (Table 5.4) to 2400 MW for an approximate cost of USD 5 billion (Energy Regulation Board of Zambia, 2019a). Paradoxically, Harrison and Whittington noted, as hydropower was being seen as a key strategy to limit impact on climate change, its availability itself may be limited by climate change (Harrison & Whittington, 2002). Their model illustrated that the Batoka Gorge hydropower scheme would be sensitive to reductions in river flows (Harrison & Whittington, 2002). In 2011, Yamba et al. used General Circulation Models (GCMs) to generate projected precipitation along the Zambezi River Basin. They found that hydropower potential will reduce for all existing and proposed schemes across sub-basins, and therefore recommended regional integration of hydropower infrastructure to include natural gas and biomass sources and for pumped storage schemes to accompany select hydropower stations to ensure continuous energy (Yamba et al., 2011). In 2012, Beilfuss highlighted that the Zambezi River Basin had one of the most variable climates of any major river basin in the world, and that the severe 1991/92 drought reduced hydropower generation and resulted of GDP and jobs (Beilfuss, 2012). He recommended prioritising investments that increased climate resilience, rather than investments into hydropower which reduced it (Beilfuss, 2012, p. 4), and investing in rehabilitation, refurbishment and renovation before investing in new infrastructure (Beilfuss, 2012, p. 36). In 2014, Spalding-Fecher et al. added to the literature on the Batoka Gorge hydropower scheme their doubts online (and in print in 2016) that it would reach its anticipated production levels from its original feasibility study (Spalding-Fecher et al., 2016).

Contrasting these highly pessimistic outlooks were the more muted results of Beck and Bernauer in 2011. One of the two supply-side scenarios they derived from the 2000 IPCC Special Report Emissions Scenarios A2 scenario projected slightly increased precipitation by 11% (Beck & Bernauer, 2011), while the other projected a basin-wide decrease in annual precipitation of 28%.[10] However, when combining these supply-side

[10] The A2 scenario is a high population growth scenario in which the population is 15 billion by 2100 (IPCC, 2000, p. 5). It assumes a significant decline in fertility for most regions and stabilisation at above-replacement levels. Economic development is primarily regionally oriented and per capita economic growth and technological change are more fragmented

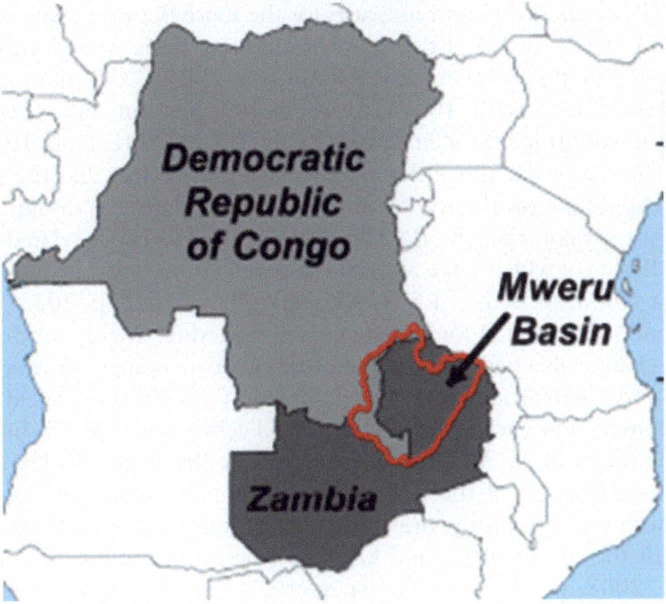

Fig. 5.3 Location of the drainage basin feeding Lake Mweru, Zambia. (*Source*: Waylen et al., 2019)

scenarios with minor, moderate and major changes in demand scenarios, not even in their minor changes in demand scenario did water availability at the end of each sub-basin in the Zambezi River Basin's 13 sub-basins increase in 2050 relative to 2000. In the minor changes in demand and supply scenario, water availability remained the same in three sub-basins—Luanginga, Lungue Bungo and Kabompo sub-basins (Beck & Bernauer, 2011, table 1) and decreased by less than 1% in the Luangwa and Barotse sub-basins. Under their 'moderate demand and supply-side changes' scenario, water availability in the Kabompo sub-basin decreased by 15%. Under their 'strong demand and supply-side changes' scenario, water availability in Kabompo sub-basin decreased by 90% (Fig. 5.3).

and slower than in other scenarios (IPCC, 2000, p. 5). Net anthropogenic CO_2 emissions from land-use change remain positive through 2100 (IPCC, 2000, p. 6). Final energy intensity in 2050 is 9.5 MJ/US$ and share of coal in primary energy by 2050 is 30% (IPCC, 2000, p. 15).

The IPCC offers different forecasts for the Zaire/Congo River Basin in 2001 and 2018. In contrast to its projections of decreased precipitation and runoff for the Zambezi River Basin, the 2001 IPCC Summary for Policymakers forecast a 10% increase in precipitation and a 10–15% increase in runoff for the Zaire River Basin (IPCC, 2001, table 10–1) by 2050. The reason for this difference would be the effect that the Indian Ocean Dipole has on the river basin, which it similarly has on equatorial East Africa. By contrast, the IPCC in 2018 limited forecasts related to the Congo Basin (of which Lake Mweru is a part), citing only one study projecting a 10% decrease in a 1.5 °C scenario (IPCC, 2018, p. 203).[11]

Beyond the potential for climate change to reduce energy production, climate change also had the potential to exacerbate competition between energy and food production. The interaction creates a so-called food, water, climate and energy nexus (Spalding-Fecher et al., 2016), bringing forth the threat of the Malthusian trap (Beck & Bernauer, 2011).

The construction of the Kafue and Kariba dams has been associated with a decline in fish production and aquatic biodiversity, and been associated with the rise of tilapia and kapenta fisheries in the Kariba reservoir (SADC, 2008, p. 12). The Kariba reservoir has resulted in damage to freshwater and marine shrimp fisheries downstream in the Zambezi Delta in Mozambique (SADC, 2008, p. 12).

Consumptive water use on the Zambezi River Basin was estimated at around 15–20% of runoff in 2007, and the Southern African Development Community (SADC) estimated that this might increase to 40% by 2025 (SADC, 2008, p. xiii; Beck & Bernauer, 2011). The largest consumptive water users of the Zambezi River Basin were dams at circa 13 km^3 per annum from evaporation impoundment (16% of consumption), followed by irrigated agriculture at circa 1.5 km^3 per annum (1.4% of consumption) (SADC, 2008, p. xiii; Beck & Bernauer, 2011). By 2025, SADC's consultants forecast that under a mega growth scenario for hydropower, hydropower would account for almost a quarter of consumptive water use, and irrigated agriculture for 6% of total use (SADC, 2008, table 2.3). Livestock would account for less than 0.2%, mining for 0.4%, industrial consumption for less than 0.1%, and urban domestic consumption for 0.65% (SADC, 2008, table 2.3)—16 times more than rural domestic consumption, in spite of being less than slightly less than half the population

[11] It is unclear to what this time horizon applies (Döll et al., 2018; IPCC, 2018).

(World Bank, 2020c). Flood releases would account for more than 6% of consumptive use (SADC, 2008, table 2.3).

Beck and Bernauer predicted that national shares of consumptive water demand would be driven mainly by population growth, which was highest in Malawi, but that Zambia would remain the biggest water user of the system and its consumption would also grow as a result of massive growth of irrigated agriculture from 34% in 2000 to up to 42% by 2050 (Beck & Bernauer, 2011, p. 1070). To provide context, according to the FAO Aquastat's database, 30% of Zambia's agricultural area had potential for irrigation by 2002, and 100% of land equipped for irrigation actually irrigated (FAO, 2020)—most Zambian farming, which accounts for 54% of the workforce (World Bank, 2020c) on subsistence farms, is rainfed (Resnick & Thurlow, 2014). This means that Zambia had substantial potential for irrigation expansion. Other countries sharing in the Zambezi River Basin's water are Angola, Botswana, Mozambique, Namibia, Tanzania and Zimbabwe. In their projections, they foresee Mozambique (downstream) and Zimbabwe (contiguous) will challenge Zambia for water consumption as all three countries are likely to experience a large decrease in water availability (Beck & Bernauer, 2011, pp. 1070–1071).

Spalding-Fecher et al. found that accelerating economic growth was increasing the potential for competition for water between hydropower and irrigated agriculture in the Zambezi River Basin, and that climate change would add additional stress to the system. Using the Water Evaluation and Planning (WEAP) tool, they assessed the Kariba hydropower dam to be vulnerable to increases in upstream irrigation demand as well as to climate change. They found that electricity generation could reduce by 12% and that expansion of the Kariba hydropower dam would unlikely deliver increases in power production even under a favourable climate (Spalding-Fecher et al., 2016). They concluded in their paper first published online in 2014 (i.e. before the power outages of 2015 and 2016) that an intense El Niño drought that affects the whole of Zambezi River Basin would result in a corresponding decline in generating capacity across the basin. Because Global Circulation Models could not model the frequency and intensity of future El Niño events, they recommended policy makers and project developers must plan for a wide variety of potential climate futures. To coordinate investment and operational decisions as effectively as possible, they recommended an energy system model for the region that would model the energy, water and agriculture economic trade-offs (Spalding-Fecher et al., 2016) and incorporate projections from

Global Circulation Models that are better at simulating the frequency and intensity of El Niño droughts.[12]

For some farms that were irrigated but did not power their irrigation using biomass, industrial farming was doubly impacted by the El Niño events: both directly from lack of rainfed water and by lack of power for irrigation. Zambia Sugar Plc and Zambeef reported reduced crop yields in 2016 due to power interruptions which affected irrigation (Zambeef Products, 2017, p. 23; Zambia Sugar, 2017, p. 36).

The above review of data and literature shows that Zambia's system of energy provision had access to an abundant wealth of information as to why it should seek to diversify from hydropower prior to the power outages to 2015 and 2016. Experience of power outages should not have been required to make the system of provision react, and yet that is what happened, as illustrated by President Edgar Lungu's comment, 'The difficult realities of climate change have taught us over-reliance on hydropower won't just do' (Mfula, 2016). The next section will delve into what alternatives Zambia's system of provision has.

5.4 Alternative Energy Sources

With the capacity utilisation of hydropower under threat by El Niño events, particularly in the Zambezi River Basin, Zambia's system of energy provision needs to consider alternatives and complements not just to hydropower in the Zambezi River Basin, but hydropower per se.

In the preceding section, we saw constructive suggestions emerge: pumped storage electricity generation for hydropower; an energy system model for the region that would model the energy, water and agriculture economic trade-offs; incorporating projections from Global Circulation Models that are better at simulating the frequency and intensity of El Niño droughts; coordination with the Southern African Development Community, the Southern African Power Pool and trade of agricultural products be used to enhance water, energy and food security in southern Africa (Yamba et al., 2011; Conway et al., 2015; Spalding-Fecher et al., 2016).

Linking regional Power Pools could achieve greater operational efficiency and hazard mitigation. Research published subsequent to the power outages

[12] In terms of forecasting future forecasts, Jury found that the most influential predictor of maize yield, river flows and GDP in South Africa was cloud depth in the tropical Indian Ocean in the preceding spring (September-November). Reduced monsoon convention is related to enhanced rain fall over South Africa the following summer and greater economic prosperity during the subsequent year (Jury, 2002).

of 2015 and 2016 indicates the need of further research to explore the implications for hydropower reliability in eastern Africa (Conway et al., 2017), whose hydropower generation may prove to be sufficient diversification from El Niño affected southern Africa. Building on this research, if interannual rainfall variability is different enough between eastern, southern and central Africa, trading across regional Power Pools (but not within) could offset the tendency at regional scales for coincident surplus supply or demand due to high reliance on hydropower production in single river basins or clusters of shared hydropower production (Conway et al., 2017). Zambia, for instance, is not diversified with respect to rainfall through its membership of the power-sharing Southern African Power Pool. According to data collected by the Southern African Power Pool, Angola, Botswana, the DRC, Malawi, South Africa, Tanzania and Zimbabwe all had installed capacities greater than their peak demand, and yet they all experienced capacity shortfalls in 2016 (Energy Regulation Board of Zambia, 2017a, p. 24). The Southern African Power Pool experienced a shortfall of 7857 MW in 2015 (Energy Regulation Board of Zambia, 2016a, p. 28) and a shortfall of 5583 MW in 2016 (Energy Regulation Board of Zambia, 2017a, p. 24). With significant drying across Africa's major basins (Niger, Nile, Sanaga and Volta), however, Falchetta et al. seem to suggest a diversification away from hydropower technology altogether (Falchetta et al., 2019).

Power can be generated in three main ways: i. through thermal generation, i.e. fossil fuel-fired plants and nuclear units, but also geothermal and biomass power generation, or concentrated solar power; ii. kinetically, including hydropower and pumped storage, wind turbines, or tidal energy; and iii. through solar photovoltaic units (Falchetta et al., 2019). All three of these main ways and several of these sub-means appear to be available to some degree in Zambia. Table 5.4 based on ZESCO's power generation expansion plan as of 2017 shows coal, hydro, solar (PV) and wind were sources for planned new power plants. Diesel and high fuel oil power plants have also been contributing to ZESCO's recent power generation (Energy Regulation Board of Zambia, 2019a, p. 34). Nuclear power is another option considered (Ministry of National Development Planning, 2017), as are waste to energy and geothermal power (Ministry of Energy, 2018).

As discussed earlier, when ZESCO and the World Bank evaluated alternatives to the Kariba and Kafue hydropower plants, the only apparent criteria that they considered was financial cost (IBRD and IDA, 1970, 1973). Since then, consultancies have formulated frameworks for evaluating the sustainability of infrastructure projects from multiple perspectives. Two that are affiliated with the UK's Institution of Civil Engineers (ICE)

Box 5.1 Criteria by which Arup's ASPIRE and the Building Research Establishment's CEEQUAL assess infrastructure projects
Arup's ASPIRE—

- Institutions
  - Structures, skills, policies, reporting
- Environment
  - Air, land, water, biodiversity, energy, materials
- Economics
  - Equity, livelihoods, macro, viability
- Society
  - Population, culture, stakeholders, services, health, vulnerability

Building Research Establishment's CEEQUAL—

1. Management
2. Resilience
3. Communities and stakeholders
4. Land use and ecology
5. Landscape and historic environment
6. Pollution
7. Resources
8. Materials, including waste
9. Energy and carbon (operational)
10. Energy and carbon (construction)
11. Water use
12. Transport

Source: Author

are Arup's SpeAR and ASPIRE and Building Research Establishment's CEEQUAL that bring together a number of indicators to assess the total environmental and socioeconomic impact of a project (Arup, 2017) – see Box 5.1. CEEQUAL goes further than Arup's appraisal tools by also assessing the resilience of new infrastructure projects (BRE Global, 2019).[13]

[13] Other broad criteria are management; communities and stakeholders (including consultation and engagement, wider social benefits, wider economic benefits); land use and ecology (land use and value, land contamination and remediation, protection of biodiversity, change and enhancement); landscape and historic environment; pollution; resource efficiency; and transport.

While both approaches would make for comprehensive feasibility studies, creating detailed feasibility studies goes beyond the scope of this study. Moreover, evaluating energy sources by 20 Arup or 12 BRE criteria dilutes the foci of this study. The Zambian system of energy provision also did not make available its feasibility studies for this study. Accordingly, we will have had to make do with using a more heuristic approach using the Pareto principle to construct a 'least regret' approach to make a very preliminary evaluation of Zambia's power generation options.[14]

In 2015, the UK National Grid introduced the concept of a 'least worst-regret' approach with the aim of minimising the cost implications of any decision made when there is uncertainty over the future (National Grid, 2015, p. 86). I suggest giving the following five criteria at least 80% weight for the appraisal of Zambia's possible power generation sources:

1. *Domestic availability of source.* ZESCO's first preference for an energy source would be within the boundaries of Zambia to a. minimise financial costs and b. to minimise the risks associated with power systems outside of Zambia's sovereign control, such as hold-up for political or financial reasons from parties outside of Zambia. Within Zambia, it can leverage its status as a wholly government-owned entity to get power generated where it is physically possible.
2. *Availability of the resource within the Southern African Power Pool (SAPP).* If the resource is not available in Zambia then it must be available within the Southern African Power Pool with which ZESCO has power interconnections until the SAPP or Zambia directly connect with power pools outside the El Niño region affected region.

Sourcing a significant portion of energy from a power pool comes with its risks but offers opportunities. Of West Africa, it has been estimated that USD 32 billion of investment and 23 million tonnes of fuel oil (and the commensurate amount of CO_2 equivalent emissions) (USAID, Tony Blair Institute for Global Change and Power Africa, 2019, p. 39) could be saved over 2020 to 2030 if the West African Power Pool traded efficiently (Gareth Walsh, introduction to Tony Blair Institute for Global Change's work in energy in Africa, 16 December 2019). The same principle of emissions and financial

[14] Named after the Italian economist Vilfredo Pareto, the Pareto principle states that for many outcomes, roughly 80% of consequences come from 20% of the causes. It has also been called the 80/20 rule and the law of the vital few. Here, we effectively assume that 20% of the criteria considered by Arup and BRE will account for 80% of weighting.

cost savings should hold for southern Africa as energy is traded between energy surplus and deficit countries to achieve operational efficiency and hazard mitigation (Conway et al., 2017). To ensure that the benefits of trade are maximised, national power utilities and governments would have to coordinate in their planning and disclose their expected surpluses and deficits accordingly. There would also have to exist a mechanism whereby energy suppliers would not be able to exploit the energy deficiencies of net energy consumers, either at the stage of contractual negotiations or in the operating stages, so as not to threaten net consumers' energy security. Multilateral bodies and international courts would likely need to be involved to regulate tariffs and resolve disputes in order for nationally owned utilities such as ZESCO and governments such as the Government of Zambia to become comfortable with outsourcing their energy generation needs. Alternatively, ZESCO or the Government of Zambia would need to be able to hold something of value to its supplier(s) hostage as collateral against default.

3. *Climate-resilient and reliable baseload.* As the preceding chapters have demonstrated, hydropower's climate vulnerability has threatened the growth of Zambia's manufacturing sector, adversely affected individual users and contributed to additional diesel-generated greenhouse gas emissions.
4. *Lifecycle greenhouse gas emissions.* If ZESCO does not want to contribute to its hydropower infrastructure becoming even more redundant than it is during El Niño events, it will want to minimise greenhouse gas emissions into the atmosphere to minimise the increase in temperatures that lead to the increase in frequency of El Niño events. Even if Zambia's direct contribution to a prospective increase in emissions is minimal, maintaining climate leadership theoretically provides the government more leverage in international negotiations with high emitting nations to reduce their emissions: the Government of Zambia has committed to reduce its emissions by 47% from its emissions in 2010 by 2030 (Government of Zambia, 2016). To take emissions into account will mean internalising to the investment appraisal process the negative externalities of lifecycle greenhouse gas emissions emanating from hydropower which, if excluded, would otherwise have represented a market failure (Coffman et al., 2020, p. 162). In a lifecycle assessment of greenhouse gas emissions emanating from hydropower, embedded carbon, the loss of carbon sinks in perpetuity, emissions from the

reservoir lake as well as the emissions of backup diesel generation when dependence on hydropower does not deliver sufficiently against a changing energy context (Ahmed et al., 2020) must be counted.[15,16,17]

While lifecycle emissions should be the criteria used, these values would only be known with the execution of bespoke feasibility studies, especially in the case of hydropower projects. I have therefore used operating emissions in Table 5.6.

5. *Effects of waste on human health and the environment.* CEEQUAL considers the effects of pollution (BRE Global, 2019), ASPIRE considers the effects of projects on the environment and health (Arup, 2017). Farquharson et al. considered not only the greenhouse gas emissions from diesel generation but also the nitrogen oxides (NO_x), sulphur oxides (SO_x) and carbon monoxide (CO) which both harm human health and also contribute to acid rain and the destruction of carbon sinks in the form of forests (US EPA, 2020b). To proxy for NO_x, SO_x and CO emissions in Table 5.6, I used kilogrammes of N_2O emissions per TJ of energy release for fossil fuels in table 2.2 of the IPCC Guidelines for National Greenhouse Gases (Gómez et al., 2006). For nuclear power, I considered nuclear waste.
6. Lifecycle financial cost. This is the one appraisal criterion ZESCO and the World Bank seem to have assessed Zambian power projects by in 1970 and 1973 (IBRD and IDA, 1970, p. 12, 1973, p. 12).

[15] Looking at proposed hydropower dams and the associated distribution infrastructure required in Chilean Patagonia, Mar (Mar, 2009) forecast that the carbon impact of construction equipment and machinery, transport of labour, material embedded energy, and land-use change would be 48x the impact of natural gas plants that would deliver an equivalent amount of energy.

[16] Reservoirs themselves contribute to 1.5% of global anthropogenic emissions from carbon dioxide, methane, and nitrous oxide reported by the UN Intergovernmental Panel on Climate Change (Deemer et al., 2016).

[17] Coffman et al. (2020, p. 164) suggest, beneficiaries should bear the responsibility for embedded carbon emissions (category 2 Scope 3 emissions according to the Greenhouse Gas Protocol (2013)). For hydropower, this would mean the substantial emissions associated with the manufacture of cement in concrete reservoir walls. For hydropower, the emissions associated with lake emissions should also be counted as Scope 1 operating emissions. Another type of emission should also be counted which is not captured within any of the Greenhouse Gas Protocol Scopes of emissions: foregone carbon sinks lost in perpetuity, as is the case with hydropower when swamps and forests are lost to make for hydropower reservoirs.

Table 5.6 A least-regret framework for assessing Zambia's potential energy sources

Source of energy	Domestic availability of resource	SAPP availability of resource	Climate resilient and reliable base load	Operating direct+ indirect GHG emissions – kg CO2e/GJ[66]	Harmless to health, environment? kg N2O/TJ	Low financial cost	Other considerations particular to the energy source
Hydropower	✗ Not with El Niño events	✗ Not with El Niño events	✗ Not with El Niño events	o mg 450 CO2e /sq. m/d (Deemer *et al.*, 2016, table 1)	✗ Loss of carbon sinks, loss of homes	✓ If capacity utilisation is high	Competition for water with agriculture
Coal	✓	✓	✓	✗ 95.1 + 14.55	✗ 1.5	✓	
Natural gas	✗	✓	✓	o 56.2 + 14.1	o 0.1	✓	
Nuclear	✗	✓	✓	✓	✗ Hazardous waste	✓	Magnitude of impact of mismanagement Required scale
Heavy fuel oil	Imported	Imported	✓	✗ N/A + 17.4	✗	✗	
Diesel	Imported	Imported	✓	✗ 74.4 + 14.55	✗ 0.6	✗	
Biomass	Untested quantity	Cannot be easily authenticated	Depends on availability	54.7-114 + N/A	✗ 0.1-4	Depends on resource	
Industrial waste	Untested quantity		Depends on availability	✗ 145 + N/A	✗ 1.5	?	
Solar	✓	✓	✗	✓	✓	✓	
Wind	✗	✗	✗	✓	✓	✗	
Geothermal	?	?	✓ if available	✓	Possible seismic activity	✓	

Key: ✓ = yes, × = no, o = intermediate. *Source*: original analysis

[a]Carbon dioxide emission factors for CO_2, CH_4 and N_2O sourced from the IPCC Guidelines (Gómez et al., 2006, table 2.2) and converted to CO_2e using a value of 25 global warming potential for methane and 298 for N_2O, versus 'Well to Tank' lifecycle carbon dioxide equivalent emission factors sourced from the UK Department for Business, Energy and Industrial Strategy (BEIS, 2019, table 4)

7. *Other considerations particular to the energy source*. This criterion picks up the nexus approach for hydropower and the particular hazards of nuclear energy.

Table 5.6 presents Zambia's energy sources assessed by these five criteria to the extent that data coverage allows, in a least-regret framework.

Hydropower: The drawbacks of hydropower reliability particularly in the Zambezi River Basin in the face of increasingly frequent El Niño events as global temperatures rise has already been discussed extensively in in the previous section (see Table 5.7 for the low capacity utilisation rates in 2016), as has the necessity for considering how it competes with food production domestically as well as the needs of neighbouring countries, with challenges from Zimbabwe and Mozambique forecast. Large hydropower is also not as climate-friendly as popularly perceived. First, there is the embedded carbon that goes into their concrete walls. Second, the flooding required to fill their large reservoir lakes destroy biodiversity and destroy carbon sinks such as forests and swamps whose ability to decarbonise has been lost in perpetuity. Looking at proposed hydropower dams and the associated distribution infrastructure required in Chilean Patagonia, Mar (2009) forecast that the carbon impact of construction equipment and machinery, transport of labour, material embedded energy, and land-use change would be 48× the impact of natural gas plants that would deliver an equivalent amount of energy.

Reservoirs contribute to 1.5% of global anthropogenic emissions from carbon dioxide, methane and nitrous oxide reported by the UN Intergovernmental Panel on Climate Change (Deemer et al., 2016). So high are the methane emissions associated with some reservoirs that Ocko and Hamburg (2019) found that hydropower facilities in Africa have

Table 5.7 Average capacity utilisation over 2016 of Zambia's largest hydropower dams

Average capacity utilisation of Zambia's largest hydrodams over 2016

	Kafue Gorge	Kariba North	Kariba North Ext	Victoria Falls	Aggregate	Source
Energy output in 2016, GWh	5853	2964	672	754	10,243	ERB 2017, p32
Installed capacity in 2016, MW	990	720	360	108	2178	ERB 2016, p57
MW/GW | h/y	1000	8760				
Energy output at 100% capacity utilisation, GWh	8672	6307	3154	946	19,079	
Average capacity utilisation over 2016	67%	47%	21%	80%	54%	

Author's calculation

Data source: Energy Regulation Board of Zambia (2016a, p. 57, 2017a, p. 32), original analysis

near-term climate impacts per unit energy generation that are multiple times that of coal-fired power plants. Using a model developed by Scherer and Pfister (2016) and not representing empirical measurements, they found that the reservoir for Kariba dam is 8× higher than coal emissions per unit of energy for near-term impacts and 3.5× higher than coal for long-term impacts (analysis emailed by Ocko to author on 18 March 2021). By contrast, the average of 1473 hydroelectric facilities spanning 104 countries and covering 43% of global hydropower electricity generation in 2009, emit a half of what coal power plants do per unit of energy in the near-term and a third in the long term (Ocko & Hamburg, 2019; analysis emailed by Ocko to author on 18 March 2021). The magnitude of greenhouse gas emissions from a hydropower dam reservoir depends on several factors, including submerged vegetation and soil; age, area, volume and depth of the reservoir; the extent of reservoir drawdowns that lead to fluctuating water levels; as well as temperature and precipitation. Smaller reservoir surface areas and methane capture mitigate emissions (email by Ocko to author, 12 March 2021). So does building in areas with lower temperatures (email by Ocko to author, 12 March 2021), but this cannot be planned for in Zambia, especially given the global warming scenarios where Zambia's average temperatures are expected to rise higher than the global average (IPCC, 2018).

Ahmed et al., referenced in Chap. 3, show how complete reliance on hydropower also increases greenhouse gas emissions by inducing electricity consumers to self-generate electricity using backup diesel generators (Ahmed et al., 2020).

Beyond contributing to emissions, hydropower construction has further adverse environmental and social consequences. Deforestation has also been linked with the rise in global pandemics arising from zoonotic disease such as Ebola (Olivero et al., 2017) as wildlife such as bats lose their natural habitat and move closer to human populations. Dam reservoirs destroy communities' homes, livelihoods and sacred lands, as they have in Zambia (Serpell, 2019). An estimated 57,000 BaTonga people were compulsorily displaced from the colonial British government's appropriation of the Kariba Dam and then excluded by the colonial and post-independence governments from accessing the dam's resources (fish, water for irrigation and wildlife), resulting in their socioeconomic as well as religious impoverishment (Matanzima, 2021). They were moved to 'less productive, problem-prone areas' which were 'so seriously degraded' that they resembled 'lands on the edge of the Sahara Desert' (Scudder, 1993). Any future hydropower reservoirs that displace people would require the Government of Zambia to engage affected people early on in the planning; assist those displaced overcome the social, psychological and physiological stress of moving; help them re-form communities; and assist them with integration into wider territorial, political and economic networks (Scudder & Colson, 1982; Matanzima, 2021).

Coal: Coal in Zambia so far costs USD 0.1035/kWh (Mabumba, 2017). However, because coal's capacity factor can be reliably assumed as 93% (Lazard, 2018, p. 18), compared with hydropower's capacity utilisation of 54% in 2016 (see Table 5.1), its effective tariff would be USD 0.06/kWh.[18] Coal, however, emits even more carbon dioxide than diesel generation does (Table 5.6) and so would contribute to the global warming that is contributing to the increased frequency in El Niño events that are impacting the efficacy of Zambia's hydropower plants. On its own as the largest producer of coal in Zambia, Maamba Collieries Ltd had coal reserves of 103 M tonnes of high-grade coal and 70M tonnes of low-grade coal as of 2012 (Maamba Collieries, 2012). This is enough to generate almost 1400 TWh of energy. Even if only half of the high-grade coal was now available, this would be enough to meet Zambia's projected grid 2030 electricity demand of 18,700 GWh per annum for 23 millennia.[19]

Natural gas: As a dispatchable and dependable energy source emitting approximately 60% of the level of CO_2 emissions that coal does during combustion, nature gas seems at first blush to be a more climate friendly baseload solution than coal. With major gas discoveries made in Tanzania and Mozambique, the resource is accessible through the Southern African Power Pool as both nations are members of the power pool. Indeed, 40% of global gas discoveries have been made in Africa between 2011 and 2018 (IEA, 2020a), and Tanzania and Mozambique have 57 and 180 trillion cubic feet of proven resources (Ford, 2016). However, major methane leakages are associated with the supply chain of natural gas (Marchese & Zimmerle, 2018), the global warming potential of which is 28–36 carbon dioxide equivalents (CO_2e) (US EPA, 2020a). When one also considers the effect of other noxious gases on human health and forests, however, choosing natural gas over coal is unambiguously justified: coal emits 15 times more nitrous dioxide (Gómez et al., 2006, table 2.2).

[18] USc 10.35/(93%/54%).

[19] One kilogram of coal generates approximately 8 kWh of energy (Ovo Energy, 2018).

Nuclear: Like natural gas, nuclear power generation is theoretically available through the Southern African Power Pool. The 1940 MW Koelberg nuclear power station in South Africa is the only nuclear power plant in Africa. Nuclear power plants produce no emissions during operation. Nuclear power once installed is cheap and a reliable source of peak demand energy. Uranium is available in South Africa, Namibia and Niger (Teljeur et al., 2017, p. 205).

However, nuclear power is complicated. First, it requires a relatively large scale power plant, above 900 MW of installed capacity, technical expertise and significant upfront cost (Teljeur et al., 2017, p. 205). Second, it can be dangerous. Within three decades, a superpower and an OECD country failed to prevent accidents at nuclear power plants due to incompetence (Chernobyl) and due to a natural disaster (Fukushima). To be more resilient to natural disasters, it seems, nuclear power needs to be located away from tectonic fault-lines and away from the ocean. Further, Pravalie and Bandoc (2018) assert that there is apparent global lack of interest in hazardous waste management, considering that underground facilities for relatively safe storage have yet to be built. Given the health and safety risks associated with managing nuclear power and waste safely and the requisite strong systems of governance, Zambians that I talked with[20] did not feel comfortable with the idea of nuclear power being managed by ZESCO, which has proven unreliable in communicating accurate load shedding schedules. That would make reliance on South Africa's power utility Eskom for nuclear power necessary if nuclear power is to be an option, given that Eskom is the only African utility with nuclear expertise. Even with its relatively positive track record, however, Eskom and South Africa in general do not exist in a stable institutional environment with good governance. Radio France International reports that plans to procure a new nuclear power plant made between former president Jacob Zuma and agents of Russia were 'secured with hefty bribes' (Cornish, 2019). Given that Zambia requires additional energy and the prospect of procuring additional nuclear power through good governance and with a stable institutional environment through the SAPP looks currently unlikely, nuclear power does not seem to an option at this point.

Diesel and heavy fuel oil: Neither diesel nor heavy fuel oil is low cost nor low in greenhouse gas or harmful to health emissions. This perhaps

[20] Interview with Chipego Zulu, CEO of Zambia Association of Manufacturers on 28 May 2018.

explains why ZESCO does not plan to expand its energy generation capacity with these fuels.

Biomass: The annual reports of Zambia Sugar revealed that the company has been generating electricity from combusting biomass to reduce its grid electricity demand and increase its resilience through periods of load shedding (2016, pp. 1, 11–12, 20, 2017, pp. 24, 41, 2018, pp. 8, 31). Biomass is technically renewable so long as the fuel source is replenished as quickly as it is consumed, but when this ceases to be the case, it is no longer renewable. From an environmental perspective, it may be challenging for the national utility to source enough renewable resources for nationally owned power plants, in a way that an agricultural company that supplements its grid electricity demand with its own waste product does not find it so challenging. And though renewable, unlike intermittent renewable sources, biomass emits greenhouse gases as well as gases harmful to health, particularly when solid fuels are burnt (Gómez et al., 2006, table 2.2). The volume of emissions associated with biomass can be worse than those by diesel generators. When operators run out of biomass, they can resort to petrochemicals, such as used tyres. For these reasons, biomass is not an environmentally attractive solution (Gibbs, 2019).

Industrial waste: Industrial waste is another possible source of energy, but again more at the company level than at a national utility scale given the unpredictable supply of fuel. Companies such as KPA Unicon construct 5–40 MW independent power plants for manufacturing firms, fuelled by those firms' waste, such as steel mill blast furnace gas or paper mill pulp (personal interview on 2 November 2018; Ahmed, 2021). The high level of GHG and harmful to health emissions also do not make them attractive for nationally scaled projects.

Solar power: Solar power is intermittent and therefore is not a reliable and dispatchable source of energy without batteries. It should not account for more than 15% of grid energy according to an energy expert with three decades of gird experience (personal communication with an energy expert at a donor agency, 27 February 2019; Ahmed, 2021). Department of Energy officers said (personal communication, 6 June 2018; Ahmed, 2021) that ZESCO's planned solar plants would not have batteries to store the energy they generated, nor would they be used to recharge the hydropower reservoirs by powering pumps. Instead, they would feed directly into ZESCO's grid. In times of high cloud cover, pollution or ash cover from volcanic activity, the sun's rays would only faintly reach the solar panels for a few hours, and in times of total solar eclipses, not at all.

The average irradiation would vary by season. It cannot therefore be relied upon for baseload energy, even if it can substitute away from a small portion of coal or hydropower during baseload hours. There is also a mismatch between the time when solar irradiation peaks and the time when demand for energy peaks in Zambia.

ZESCO charges peak tariffs between 18:00 and 22:00 (Energy Regulation Board of Zambia, 2017b, p. 6). Peak tariffs occur when there is peak demand for power, as people return home from work and turn on their house lights as the evening begins and switch on their electrical appliances. As the graph below shows, however, solar irradiation peaks at about midday. In a sample of four days four months apart, there was only solar irradiation on 27 February 2017 at 18:00, and none on the other three days. There was no solar irradiation any of these days at 19:00, 20:00, 21:00 or 22:00, which is when ZESCO's peak tariff ends (Fig. 5.4).

Given all these plans and limitations, solar would be most useful for ZESCO in decreasing the variable charge it pays for thermal power, or, when hydropower reservoirs are low, conserving limited hydropower water reserves during daylight hours. To maximise its usefulness, however, Zambia's system of energy provision would do well to change their plan to not use solar to recharge hydropower reservoirs.

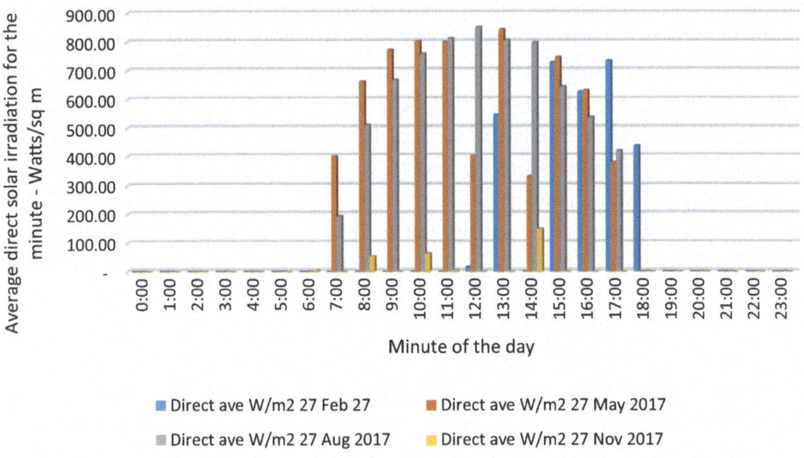

Fig. 5.4 Average solar irradiation for a minute every hour of four days each four months apart at Lusaka. (*Source*: Generated from data from the World Bank, 2017b; graph: own)

Solar storage: In the medium-term, repurposed second-life electric vehicle lithium-ion batteries could offer a low-cost storage solution that could pair with solar power plants to offer dispatchable and reliable power (Global Battery Alliance, 2019; McKinsey, 2019). However, it remains to be seen whether there will be enough global supply of lithium-ion batteries to help bridge the gap in energy demand. The supply of electric vehicle batteries will be concentrated in China and Europe, where vehicle emissions-reduction regulation, concessional credit and battery recharging infrastructure for electric vehicles will drive demand for electric vehicles (IEA, 2020b).

Estimates for the potential supply of electric vehicle reused batteries available for stationery use vary largely. The most aggressive forecasts, such as McKinsey's breakthrough case (McKinsey, 2019), suggest that the installed capacity of lithium-ion batteries available for second-life use could be twice as great as what Vivid Economics (2020) forecast would be the universal energy access demand for installed power capacity in Africa by 2030. McKinsey's base-case scenario and the International Energy Agency's forecasts suggest that it could amount to approximately meet that demand (McKinsey, 2019; IEA, 2020b; Vivid Economics and Oeko-Institut, 2020). Tempering expectations that such supply will be available to lower income countries, however, are the prospects that batteries will be used for longer than they have been projected to in their first lives as vehicle batteries.

Whatever the global supply, domestic demand and transport costs and logistical complications will reduce the availability of lithium-ion batteries available for use in Africa. African demand for Chinese and European electric vehicle batteries will compete against domestic demand for stationery applications, such as telecommunication towers and data centres (Circular Energy Storage, 2019; Willing, 2020). Additionally, importing used lithium-ion batteries to Africa will involve the additional costs of attaining UN38-3 certification, the challenge of successfully procuring refrigerated shipping containers in order to avoid thermal runaway and the potential for fire; and the financial and environmental costs of using energy-intensive refrigerated containers. These are already in limited supply even before the anticipated future exponential growth in supply of second-life batteries. The Head of Energy Storage Business of a multinational battery company reported missing a shipping deadline by two months because of lack of refrigerated containers (personal interview, 15 December 2020).

This same enabling policy and infrastructure for electric vehicles that exists in China and Europe is lacking in Africa in general and in Zambia in particular. At most, based on DNV GL (2020) forecasts of electric vehicle battery supply in sub-Saharan Africa for every year between 2020 and 2030, if the lifespan of two-wheeler and automobile batteries was five and ten years respectively, the installed capacity of electric vehicle batteries available for repurposing for stationery grid applications would be 2 GW in 2030 across sub-Saharan Africa. Lifespans could be substantially longer, however, depending on appetite for reduced storage capacity in first life batteries, especially among second-hand electric vehicle owners.

Wind: Wind energy potential in Zambia is 'relatively limited' and not expected to attain high capacity factors (Africa Power Ventures, 2018, p. 7).

Geothermal: At present it appears that geothermal energy is not a serious contender for providing a significant portion of Zambia's energy needs. Geothermal energy is physically reliable because it is consistent, efficient, and can easily accommodate changes in electricity demand (Lofthouse et al., 2015, p. 1). Geothermal plants produce few emissions and are more environmentally friendly than traditional fossil fuel plants (Lofthouse et al., 2015, p. 1). However, the question of availability of geothermal power in Zambia is yet unproven. The costs of exploration involve high capital expenditures for drilling and plant installation and the returns are uncertain. As of 8th August 2020, the latest news on exploration for geothermal power in Zambia is that the developer Kalahari GeoEnergy Ltd has raised USD 3.2 million in the form of a convertible loan to drill wells at Bweengwa. While the ultimate goal is to construct a 15 MW steam power plant, the second part of the loan is conditioned on one of the wells demonstrating a deep reservoir temperature of at least 130 °C (Richter, 2020; Takouleu, 2020). An environmental and health downside of geothermal drilling is that it can induce earthquakes (Harmon, 2009).

From this preliminary analysis of Zambia's alternative sources of energy, it seems apparent that gas-fired power will be Zambia's least-regret source of energy in the short-term in terms of resource availability, dispatchability, greenhouse gas emissions and financial cost, if ZESCO and the Government of Zambia can become comfortable with relying upon neighbouring TANESCO and Electricidade de Moçambique as suppliers. If the governments and energy regulators of Tanzania andMozambique allow, ZESCO could even contract directly with independent power producers in those countries for power, and pay wheeling charges to the utilities for using their distribution infrastructure. On top of this, Zambia's system of energy provision should work to connect with utilities and power projects beyond those in southern Africa which do not share its weather patterns. In the medium-term, the prospect of dispatchable solar power with the use of affordable battery storage if not recharged hydropower looks like it could deliver an emission-free and reliable source of energy. The high embedded cost of carbon in photovoltaic cells (Gibbs, 2019) and the issues of energy intensive recycling or toxic landfills would be new environmental challenges to overcome with end-of-life batteries (Boyden et al., 2016).

Having seen that Zambia has alternatives to hydropower, we can now grapple with why Zambia's system of energy provision did not seek a balanced energy portfolio mix.

5.5 The Historical Institutional Path Dependence Framework

Path dependence refers to institutional norms that become ossified at a point in time—a critical juncture—through positive feedback. It is the 'notion that, for any given sequence of practices and events, past choices and temporally remote events can help explain subsequent paths of institutionalised practices and contemporary outcomes' (Colignon, 1997, p. 39). Mahoney explains (2000, pp. 513–14) that in a path-dependent pattern, selection will happen in an otherwise unpredictable manner other than what has been tried and tested before.

Adapting and expanding on Mahoney's explanation, at Stage 1, there exists a set of suboptimal initial conditions at which point a universe of interventions A, B and C could be adopted.[21] Theory is unable to predict which will be chosen (2000, p. 538). Option B is chosen at Stage 2. Stage 2 is described as a 'critical juncture'—a particular point in time where a decision is made. At Stage 3, the past choice triggers the feedback mechanism that informs and hence limits future decisions to repeat the same decision taken at Stage 4 and beyond.

[21] In his illustration of contingency of a self-reinforcing sequence, Mahoney (2000, Fig. 1) shows three distinct times, as opposed to stages. He does not give reflection of Time 2 its own distinct time period, as I do.

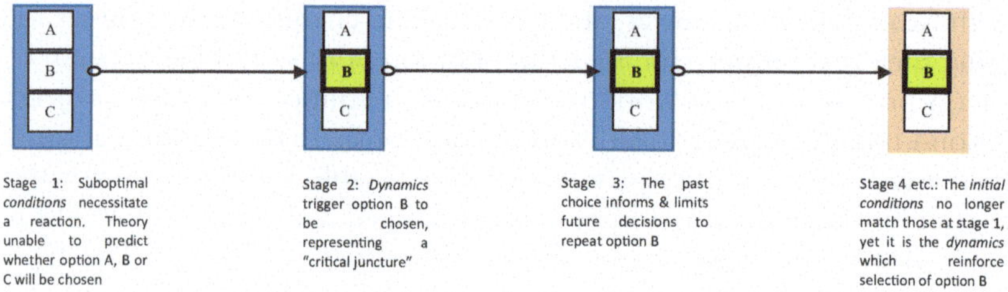

Fig. 5.5 Path dependence illustrated. (*Source*: Ahmed, 2020)

At Stage 4, Pierson and Skocpol (2002, p. 700, with reference to Stinchcombe, 1968) would argue that it is the *dynamics* triggered at the critical juncture, rather than the initial conditions themselves, that cause reproduction of those same dynamics. Selection of Options A and C becomes increasingly unlikely, although not impossible (Fig. 5.5).

In this work, we adapt the institutional path dependence framework to organisations within Zambian energy's system of provision.[22] Douglass North wrote that organisations are created to take advantage of the opportunities that institutions determine. 'The resultant path of institutional change is shaped by the lock-in that comes from the symbiotic relationship between institutions and organisations that have evolved as a consequence of the incentive structure provided by those institutions and the feedback process by which human beings perceive and react to changes in the opportunity set' (North, 1990, 5%). Institutionalisation occurs, according to Colignon (1997, p. 39), 'when organisational structure and practices—which maintain relationships and upon which policies are derived, rest upon, and sustained—are themselves, maintained through embeddedness in path-dependent relations.'

Taking forward the concept of lock-in in an organisational theory context, or 'structural inertia' (Cecere et al., 2014, p. 1042), Cecere et al.'s broad literature review of multiple academics (Dosi, 1982; Nelson & Winter, 1982; Tushman & Anderson, 1986; Levitt & March, 1988; Hannan & Freeman, 1989; Henderson & Clark, 1990; Dosi et al., 1992; Christensen, 1997; Tripsas & Gavetti, 2000; Davies & Brady, 2000; Sinclair-Desgagné & Soubeyran, 2000; Becker et al., 2005) concludes the following narrative, with some additional insights added where indicated

[22] The next section discusses Zambian energy's system of provision.

with citations of the original works. Organisations are endowed with routines and competences that define and bound their behaviour and strategies, limiting their adaptive intelligence. Individual agents use routines, which can be considered 'organisational memory', to economise on cognitive resources and to make up for their bounded rationality. Routinisation entails satisficing behaviour and constitutes a rational defence against time scarcity (Sinclair-Desgagné & Soubeyran, 2000). Routines are a crucial part of any account of how organisations accomplish their tasks in society (Becker et al., 2005). For an aspect of organisational behaviour to emerge and persist as a 'routine', there must be a certain amount of stability to the conditions moulding behaviour: broad policies set by management, the conditions must be consistent with organisational goals and policies (Becker et al., 2005). Routines are thought to be the repository of organisational capabilities, but also pave the way for deliberate learning (Becker et al., 2005) inside organisations—in the Zambian energy system of provision's case, these would have resulted in increasing returns to specialised knowledge in building and operating hydropower. Developed over time along a specific technological trajectory, routines become a source of lock-in which in the face of radical and disruptive change (in this context of this study, the disruptive change is the increasing frequency of El Niño droughts) become a constraining force to adaption to the new context. Core competencies turn into core rigidities.

Adding to this analysis on increasing returns of lock-in to learning, Douglass North has argued that lock-in will attempt to reduce uncertainty and minimise transaction costs by shaping the direction of the acquisition of knowledge and skills. This direction will determine long-term development (North, 1990, pp. 42–44). 'The way in which knowledge develops influences the perceptions people have [...] and hence the way in which they rationalise, explain and justify' (North, 1990, p. 43). This does not, however, result in adaptive efficiency, concerned with inducing innovation, undertaking risk and resolving problems and bottlenecks—it is adaptively efficient institutional frameworks that permit the maximum generation of trials that will most likely solve problems (North, 1990, pp. 45–46).

Many rationale-choice theorising approaches fall short in explaining organisational rigidities because they assume that institutions are patterns of regularised behaviour that reflect Pareto-optimal equilibria to collective actions that are stable because the actors cannot improve their positions by defecting from the pattern of behaviour (Hall, 2010). Even in a world of voluntary decisions and individual maximising behaviour, the rigid 'lock-in' that path dependence creates results in inefficiencies (Liebowitz & Margolis, 1995).

Liebowitz and Margolis (Liebowitz & Margolis, 1995) distinguished three types of path dependence by the type of inefficiency they create or do not create:

1. 'First-degree path dependence', where sensitivity to starting points exist but does not result in inefficiency;
2. 'second-degree path dependence', where the inferiority of a chosen path is unknowable at the time a choice was made, and which leads to outcomes that are regrettable and costly to change; and
3. 'third-degree path dependence', where sensitive dependence on initial conditions leads to an outcome that is inefficient and was avoidable but is also remediable.

In the context of Zambia, the World Bank's energy investments in the 1950s to 1970s solely in Zambian hydropower would be second-degree path dependence whereas ZESCO's investment decisions regarding the Maamba Collieries coal-fired power plant, its current investment plans dominated by hydropower and its subsidised tariffs as part of industrial policy but which prevent it from expanding power generation are third-degree path dependence. It is this third-degree path dependence to which we will now turn.

5.6 ZESCO's Third-Degree Path Dependence

When ZESCO evaluated both the Kariba North project and the Kafue hydroelectric projects, it compared them against the thermal alternative, a coal-fired power plant at Maamba Collieries (IBRD and IDA, 1970, p. 12, 1973, p. 12). However it considered the thermal alternative on only one criterion: lifecycle cost. Maamba was evaluated to be more expensive on both a capital and an operational expenditures basis (IBRD and IDA, 1970, p. 12, 1973, p. 12). Given that the effect of El Niño Southern Oscillation events on an unbalanced energy portfolio mix four decades hence would have been unknowable in the 1970s, evaluation of the options on just this one criterion was a case of second-degree path dependence stemming from the Bank's established investment appraisal process discussed in Sect. 2.3.3. However, it seems that little changed in ZESCO's

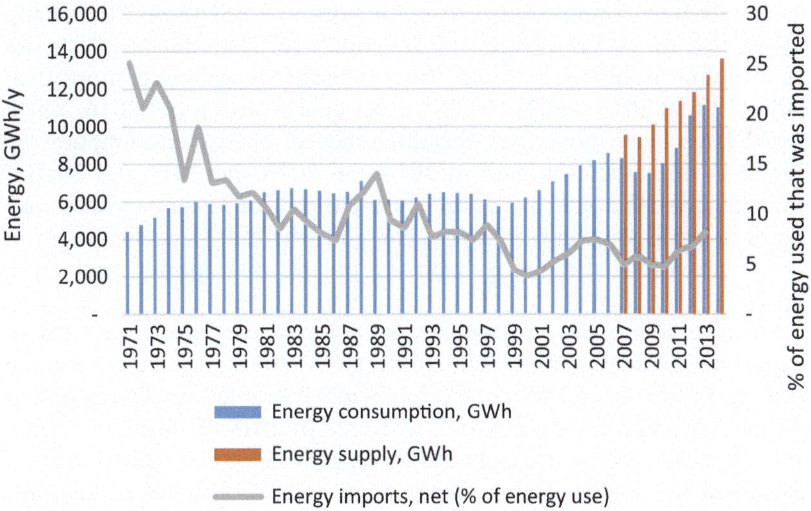

Fig. 5.6 Domestic energy consumption in Zambia versus percentage of energy imported into Zambia, 1971–2014. (Author's calculation. *Sources*: Energy Regulation Board, 2011; Energy Regulation Board of Zambia, 2015; World Bank, 2019. The energy consumption was calculated from the energy consumption per capita and population figures given)

evaluation processes from the 1970s to the 2000s such that when the inferiority of a chosen path was knowable, an inefficient outcome that was avoidable arose because of third-degree path dependence.

By 2007, ZESCO's energy supply was not far exceeding demand, as Fig. 5.6 shows. Additionally, Zambia had seen lower precipitation levels in 1992, 1994, 1995, 1999 and in 2005 than what it was to experience in 2015, as Fig. 5.1 illustrates. Additionally, IPCC pessimistic projections of rainfall in the Zambezi River Basin and Harrison and Whittington's pessimistic evaluation of the Batoka Gorge hydropower project had been published (IPCC, 2001; Harrison & Whittington, 2002) as noted earlier in Sect. 5.3. At the same time, the coal dumps, which were the by-product of mining at the Maamba collieries, were at risk of spontaneous combustion (African Energy, 2014). If Maamba Collieries had been considered since 1970 as an option for power generation, given that rainfall patterns were not reliable even in the 1990s, and given that demand was catching up with power supply, why was the Maamba Collieries coal fire power plant only fully commissioned to sell energy to ZESCO in 2017 (Patil, 2017), after the power outages of 2015 and 2016? Had Maamba Collieries come online by 2015, it could have theoretically delivered more than 2400 GWh in 2015 and 2120 GWh more energy than it did 2016, which would have been more than the difference in energy consumption by ZESCO's domestic customers in 2014 and 2015 and 2014 and 2016.[23] Why did ZESCO not act with urgency to ensure that Maamba was delivering to the grid as soon as was possible, and why did Zambia's system of energy provision continue to plan for greater hydropower generation capacity?

Maamba was again on Zambia's energy system of provision's list of prospective generation projects in 2008 (JICA, 2008), was included in the Ministry of Energy and Water Development's Power System Development Master Plan and was expected to be delivered by 2014 (Ministry of Energy and Water Development, 2011) and was approved for USD 150 million of loan from the African Development Bank in 2013 (Power Technology, 2013). A director at ZESCO explained that ZESCO could not agree to the tariff being offered by the independent power project for Maamba Collieries coal-fired power plant (personal interview, 4 September 2019; Ahmed, 2021), and that originally, the IPP was meant to deliver energy directly to a mining company. When that negotiation failed, the Government of Zambia 'imposed' agreement for the power plant upon ZESCO.[24] Even so, the reported timeline of negotiations and financing arrangements suggest that Maamba Collieries could have delivered power to ZESCO's grid before the 2015 power outages had ZESCO delivered the transmission infrastructure on time to make use of power plant's output.

The Energy Regulation Board approved ZESCO's Power Purchase Agreement with Maamba Collieries in September 2011 (Sinkamba, 2011) and the Ministry of Energy and Water Development signed an Implementation Agreement with Maamba Collieries on 16 September 2011 (Maamba Collieries, 2013, p. 6). Financing for the 300 MW plant was approved two years later by the African Development Bank in October

[23] Assuming 93% capacity utilisation and generation for 24 hours every day, to relieve ZESCO's hydropower dams running low on stored water. In 2016, Maamba delivered 326 GWh of energy (Energy Regulation Board of Zambia, 2017a, p. 8). The difference in domestic energy consumption between 2014 and 2015 was 955 GWh. The difference in domestic energy consumption between 2014 and 216 was 1548 GWh (see Table 1.1).

[24] ZCCM Investment Holdings, a successor to Zambia Consolidated Copper Mines Ltd, owned by the Government of Zambia and also publicly listed (JICA, 2008, p. 109)

2013 (AfDB, 2013). The first 150 MW was projected to come online by October 2014 (Power Technology, 2013), but commissioning was delayed by a hold-up on the transmission line to link the plant to the ZESCO grid (African Energy, 2014). A statement by the Resident Director of Maamba Collieries suggests that the delay stemmed from ZESCO's end: 'We have made substantial progress in in ensuring that ZESCO would [*sic*] put in place certain important facilities to ensure that the power we will produce will be transported on the national grid' (Maamba Collieries, 2015b, p. 3). In fact, a report by Maamba Collieries indicates that ZESCO only started to prepare for its power in 2015. Entitled 'ZESCO starts power line upgrades to prepare for MCL power', the report states that ZESCO had only just signed a memorandum of understanding with a private contractor to upgrade substations and transmission lines to be able to receive the power plant's energy (Maamba Collieries, 2015b, p. 7)—well after it was meant to accept Maamba's energy. The project was expected to be completed within 15 months (Maamba Collieries, 2015b, p. 7). Maamba Collieries' subsequent quarterly newsletter announced that ZESCO was providing a temporary arrangement for power evacuation to be completed within 5 months (Maamba Collieries, 2015a, p. 13).

ZESCO's resistance to pursuing a power purchase agreement with Maamba Collieries and its lack of determination to deliver transmission infrastructure to receive the power plant's generated energy as soon as possible raises yet more questions about the constraints on ZESCO to act on better judgement:

1. ZESCO should have independently seen the imperative of adding Maamba Collieries' energy to the grid quickly. By 2015, electricity demand was 1959 MW, compared with installed capacity of 2411 MW (email sent by an energy officer at the Ministry of Energy on 30 April 2020); average capacity utilisation of Zambia's hydropower dams only needed to fall below 81% for Zambia to experience power shortages. Why did ZESCO not see the imperative of diversifying its energy generation portfolio with Maamba Collieries?
2. Why was there disagreement on the tariff?

The next two subsections will answer this two-part question that taken together explores why ZESCO failed its existing customers.

5.6.1 ZESCO Did Not Treat Diversification from Hydropower with Urgency

Institutional path dependence explains why ZESCO did not see the urgency of commissioning the Maamba Collieries coal-fired power plant to diversify its energy portfolio from climate risks.

Stage 1 of the path-dependency model represents ZESCO in 1973. The Zambian copper mining industry required more energy supply to meet its demand (IBRD and IDA, 1973, p. 1). ZESCO did not yet know whether extension of the Kafue hydroelectric power project or construction of the Maamba Collieries coal-fired power plant was the most suitable choice for energy generation, where suitability is a function of reliability and dispatchability, cost, availability, environmental impact, climate impact, human health impact and impact on water availability for food.

Stage 2 is when ZESCO carried out ex-ante investment appraisal analysis and found that the extension of the Kafue hydroelectric power project would be less costly than the Maamba Collieries coal-fired power plant. Investment appraisal based on just one criterion—least cost (IBRD and IDA, 1973, pp. 12–13)—triggered hydropower to be chosen, representing a 'critical juncture'. Other criteria were not considered. Even had they been, it is not clear that the result would have been different, thus benign first-degree path dependence could have emerged from this critical juncture had climate conditions not changed.[25]

Stage 3 is ZESCO in the late 2000s and early 2010s when the past limits and informs the present. As Zambia's demand from existing customers catches up with supply (see Fig. 5.6), the Ministry of Energy plans for and approves the Maamba coal-fired power plant, but alongside much more installed hydropower capacity. ZESCO, meanwhile, slows the process of commissioning the coal-fired power plant.

By 2015 electricity demand was 1959 MW, compared with installed capacity of 2411 MW (email sent by an energy officer at the Ministry of Energy on 30 April 2020), meaning that average capacity utilisation needed to fall below only 81% for Zambia to experience power outages. Zambia had seen lower precipitation levels in 1992, 1994, 1995, 1999 and in 2005 than what it was to experience in 2015, as Fig. 5.1 illustrates. A growing body of literature was accompanying the IPCC's pessimistic

[25] The El Niño effect would come decades later, so excessive supply of hydropower would have been dispatchable. It is not clear how hydropower and coal fire would have compared on environmental impact when the effects of flooding were taken into consideration.

projections of rainfall in the Zambezi River Basin and Harrison and Whittington's pessimistic evaluation of the Batoka Gorge hydropower project (IPCC, 2001; Harrison & Whittington, 2002) as noted in Sect. 5.3. Yamba et al, Beck and Bernauer, Beilfuss and Spalding-Fecher et al. were all to add their voices online by 2014 (Beck & Bernauer, 2011; Yamba et al., 2011; Beilfuss, 2012; Spalding-Fecher et al., 2016), with Spalding-Fecher et al. concluding that climate change and upstream development must be incorporated into both project and system expansion planning (Spalding-Fecher et al., 2016). At the same time, the coal dumps, which were the by-product of mining at the Maamba collieries, were at risk of spontaneous combustion (African Energy, 2014).

To its credit, the Zambian energy provision commissioned the Maamba coal fired project. But it also commissioned more hydropower (Itezhi Tezhi, Kafue Gorge Lower, Kariba North Bank Extension and Kabompo), and slated for the future three more large hydropower projects (Batoka Gorge, Devil's Gorge, Kalungwishi) and smaller hydropower projects (Ministry of Energy and Water Development, 2011; Tembo & Merven, 2013).

Within Zambia's system of energy provision, the Ministry of Energy, Energy Regulation Board and international donors in the form of the African Development Bank, had agreed to contract and finance the Maamba coal-fired power plant (Sinkamba, 2011; AfDB, 2013; Maamba Collieries, 2013, p. 6). It seems though that ZESCO was not on board. A director at ZESCO reported that the government (the Ministry of Energy and Energy Regulation Board) had 'imposed' the agreement on ZESCO because ZESCO found the cost-recovery tariff to be too high (personal interview, 4 September, 2019; Ahmed, 2021). ZESCO's 2016 Integrated Report gives an insight into how ZESCO sceptically views the non-zero variable cost of thermal energy: 'we remain focused on embracing cost cutting measures such as switching from costly diesel generation to cheaper and renewable energy sources' (Zesco Ltd, 2017a, p. 7). Even with the agreement imposed, it took ZESCO until 2015 to sign a memorandum of understanding with a private contractor to upgrade substations and transmission lines to transmit power from Maamba onto the national grid, causing transmission delays after the completion of construction of the power plant, as discussed above in greater detail (African Energy, 2014; Maamba Collieries, 2015a, 2015b).

ZESCO did not see climate resilience as an important investment appraisal criterion to consider historically before 2015 and therefore did not see the need to diversify its portfolio of energy generation sources. ZESCO failed to give urgency to first the development of Maamba coal fired plant and then the required infrastructure needed to off-take power from its coal-fired power plant. ZESCO prioritised new hydropower projects. Past choices informed ZESCO's investment appraisal criteria which precluded climate resilience.

Stage 4 is ZESCO in 2017. The El Niño effect brought drought to southern Africa in 2015 and 2016 (FAO, 2015, 2018; Vidal, 2016; Hao et al., 2018), rendering ZESCO's hydropower plants ineffective in meeting demand.[26] The Kariba North Bank Extension, commissioned in May 2014, produced energy in 2015 at only 37% capacity utilisation in its first full calendar year and 21% in its second year due to the drought; average capacity utilisation of Zambia's largest hydropower dams falls well below 81% in both 2015 and 2016 (see Table 5.3). The 300 MW Maamba Collieries coal-fired plant came fully online in July 2017 (Patil, 2017), after the power outages of 2015 and 2016, outages that it could have prevented had it come fully online sooner. Four more power plants, one hydropower, two solar and one fossil-fuel based, also came online soon—Itezhi-Tezhi 120 MW hydropower plant in 2015/16, a 105 MW heavy fuel oil fired thermal power plant in Ndola in 2017, and 54 MW and 34 MW solar power PV power plants in 2019 (correspondence with a Ministry of Energy officer on 30 April 2020).

ZESCO's economists (interview on 10 November 2017) believed that they had figured out how to resolve the power outages as they presented their plan to expand power capacity, but not entirely with hydropower, as they handed me Table 5.4. Some more nuanced thinking emerged in ZESCO's 2016 Integrated Report (Zesco Ltd, 2017a, p. 7) about differentiating between hydropower projects in the Zambezi River Basin and in the Lake Mweru Basin/Congo River Basin:

> We also continue to shift our dependence from water bodies in the southern part of the country which have been hardly hit by low rainfall and water levels in the past few years. To this end a number of projects are being explored in the northern part of Zambia such as the Luapula River Basin and Kalungwishi Hydro Power Projects.

[26] Also to north South America, and south and south east Asia, making this case study applicable to other hydropower dependent countries beyond southern Africa.

The call to diversify ZESCO's water catchment areas as a main source for hydropower (Zesco Ltd, 2017a, p. 7, 2018, p. 21) demonstrates that individual minds as well as at an institutional level, ZESCO had recognised that the initial conditions for investment in hydropower in the Zambezi River Basin no longer matched those at Stage 1. Yet path dependence at the institutional level ensured that Zambezi River Basin hydropower projects including the 2400 MW Batoka Gorge Project remained in ZESCO's next annual Integrated Report (2017, the most recent one available) and even in the Energy Regulation Board's 2018 Energy Sector Report (the most recent one available) (Zesco Ltd, 2018, p. 45; Energy Regulation Board of Zambia, 2019a, p. 56).

Path dependence explains ZESCO's and more broadly Zambia's system of energy provision's continued proclivity for hydropower in spite of the evidence that hydropower in an El Niño-vulnerable southern Africa is no longer the dependable, dispatchable source of hydropower that it once was capable of being when capacity for supply far exceeded demand. Having explained why Zambia's system of provision failed to diversify Zambia's power generation capabilities from Zambia's climate vulnerability, the next subsection will supplement this explanation with Zambia's system of provision's failure to achieve system sustainability.

5.6.2 ZESCO's Tariff Structure Failed to Achieve System Sustainability

The previous subsection mentioned that a director at ZESCO stated that ZESCO had difficulty agreeing to the tariff being offered by the independent power project for Maamba Collieries coal-fired power plant (personal interview, 4 September 2019; Ahmed, 2021), and it took imposition by Government of Zambia for the tariff negotiations to not breakdown. This subsection explains why it is likelier that out of the negotiating parties on tariff, ZESCO is the likelier to have held a position misaligned with the interests of its customers. It then attempts to explain why that would have been the case through two mechanisms. First, why inherently unequal distributional policy between residential customers and residential non-customers never became a cause for voter mobilisation. Second, why policy aimed at allocating low tariffs to industries that would spur the economy would result in a lock-in of low tariffs even when the system of provision required additional funds to ensure reliability of electricity delivery.

Economic theory would model the developers of Maamba Collieries, Nava Bharat, as profit-maximising investors, seeking a minimum risk-adjusted return given their options to invest elsewhere. While Nava Bharat would have been incentivised to negotiate to maximise profit by seeking to gain returns in excess of the norm for the risk of their investment, they would not have been incentivised to draw out negotiations over a long period, since this would incur extra running costs before they could hope to be compensated, because their chances of winning the commission would diminish and because the opportunity cost associated with pursuing this venture would have increased. JICA documents their interest and involvement with Maamba Collieries back to 2008 (JICA, 2008). The resulting negotiated tariff between Nava Bharat and ZESCO should not therefore have been excessive from the perspective of returns Nava Bharat could expect by investing elsewhere for a similar level of risk. Based on my own experience with a peat-fired power plant in another landlocked African country (I was a Transactions Advisor to the Rwandan power utility through the Rwanda Development Board from 2012 to 2016), the tariff of USc 10.35/kWh (Mabumba, 2017) does not seem excessive on the basis of a 15% post-tax equity internal rate of return.[27]

To answer the question from ZESCO's perspective, further context is required. In a ministerial statement to Parliament, the Minister of Energy revealed that ZESCO charges on average USc 5.47/kWh to its residential, industrial and mining customers, while it pays Maamba Collieries USc 10.35/kWh (Mabumba, 2017). This is approximately the tariff that ZESCO would need to charge to cover the cost of asset maintenance and new energy infrastructure (confidential document viewed). And yet as of 2014, ZESCO offered the lowest tariffs in sub-Saharan Africa, with the average Zambian tariff only 38% of the median one (Trimble et al. 2016). Even as far back as 1973, the World Bank noted (IBRD and IDA, 1973, p. ii):

> ZESCO's past financial position has not been satisfactory. This was largely due to low tariffs both for its retail power supplies to its consumers and for its bulk supplies of power from its generating stations to the [Central African Power Corporation] and on account of liabilities inherited from predecessor companies.

[27] On a prima facie basis, without knowing the associated costs and without knowing the structure of the tariff and therefore what proportion is escalated with inflation.

In that instance, the World Bank agreed to support ZESCO on the condition that it raised its tariffs by 25%. More recently, the World Bank estimated that the Zambian power sector lost USD 300–400 million in 2016 due to under-pricing (World Bank, 2017a, p. 5), while a Zambian newspaper published that ZESCO owed Maamba Collieries USD 149 million as of November 2019 (Diggers Editor, 2019).[28] Clearly, ZESCO's tariff policy did not make sense from a system sustainability perspective. This is why a company like Africa GreenCo sees a gap in the market for a credit-worthy off-taker in Zambia to encourage new investment by independent power producers (interviews with their CCO on 23 April 2019 and CEO on 1 August 2019 as well as with an energy expert at a donor agency on 27 February 2019; Ahmed, 2021).

ZESCO's tariff policy did not make sense from a distributional equity perspective either. Two-thirds of Zambia's population is without ZESCO electricity (ZICTA and CSO, 2018) and ZESCO's ability to expand its electrification access is compromised by subsidising existing users. This makes it unfair to those without any electricity for ZESCO to subsidise even their poorest customers. Cost-recovery tariffs across the board would enable ZESCO to invest more in grid electrification. This would help achieve universal access, to the extent that grid electrification is more cost-effective than privately offered or subsidised off-grid electrification.

Sabatier's Advocacy Coalition Framework (Sabatier, 1998) or alternatively even corruption might explain the low tariffs. However, the evidence does not support either interpretation.

First, the Minister of Energy's parliamentary address disclosed that the largest mining companies are ZESCO's highest tariff payers, paying USc 9.3/kWh (Mabumba, 2017). Given that mining companies consume more than half of ZESCO's generated electricity and that 80% of Zambia's mining output is generated by just four companies, advocacy should result in the largest, most consolidated industries benefiting the most instead of the least. Even if the Minister's statement about mining companies was inaccurate or only temporarily true (he was talking of an 'interim' tariff), my conversations with the Director of Economic Regulation at the Energy Regulation Board (personal interview, 6 June 2018; Ahmed, 2021) and an energy expert at a donor agency (phone interview, 27 February 2019)

[28] I heard similarly from a private developer on 4 September 2019 in Lusaka. I was informed that ZESCO was only paying Maamba Collieries the debt service payment portion of the capacity charge and the variable fuel cost.

indicated that mining companies receiving ZESCO energy through the Copperbelt Energy Corporation would still be paying more than residential and industrial customers, and close to the value that the Minister had stated.

There also does not seem to be an intersectoral coalition between manufacturers and mining companies, and it seems manufacturers would be open to paying higher tariffs. An interview with the CEO of the Zambia Association of Manufacturers (personal interview, 28 May 2018; Ahmed, 2021) indicated that the manufacturing sector perceives an unfair tariff distribution between sectors and would like to see a rise in tariffs for the mining sector: they would be willing to pay more if first the perceived unfairness of the mining sector not paying enough was solved. She also said that the manufacturing sector would consider agreement to further tariff increases pending the publication of the government's much delayed and awaited Cost of Service Study for the electricity sector.

Third, residential customers are not organised to the extent that the consolidated mining sector is. Yet the third of the population that is privileged enough to have grid electricity at home enjoy even lower tariffs than the manufacturing and mining sectors.

Voter concentration in urban areas can partially explain the lack of distributional equity in ZESCO's tariff structure. When Michael Sata ran to end the Movement for Multiparty Democracy's two decade rule on a campaign against inequality, he meant urban inequality (Resnick & Thurlow, 2014). In rural areas, he relied on ethno-linguistic appeals to the Bemba populations (Resnick & Thurlow, 2014). His non-ethnic constituency was the urban poor, opening and closing his campaign in Lusaka's informal market and shanty compound, rather than the rural poor (Resnick & Thurlow, 2014). In 2011 he won 63% and 68% of the vote in urban Lusaka and Copperbelt Provinces, and 64% and 73% of votes in Bembaphone Northern and Luapula Provinces. This helped him amass 43% of the Zambian national vote against the incumbents and eight other opposition parties to be declared winner of the presidential elections (Al Jazeera, 2011). His Patriotic Front party also emerged the most represented in the National Assembly with 40% seats in a first-past-the-post system. Two-thirds of urbanites now have grid access as opposed to 6% of rural households (ZICTA and CSO, 2018). Since the year of his election in 2011, urban access to electricity (both on- and off-grid) increased from 59% to 75% in 2017, while in rural areas it increased from 5% to 14% (World Bank, 2020c). An issue-based, needs-based appeal to electrify the two-thirds of the rural population without grid electricity would prove more difficult for the same reason that infrastructure access to rural areas proves more difficult to justify: fewer people are reached with the same fixed costs.

Having explained why unequal distributional policy between residential customers and residential non-customers has not yet become a cause for voter mobilisation, we can now examine the industrial policy that informed the choice of low tariffs in the first place that were to become locked-in and prevent ZESCO from imagining paying USc 10.35/kWh for thermal energy.

A path-dependent belief in the allocative efficiency of charging subsidised tariffs offers a credible explanation for why ZESCO became married to under-pricing tariffs and used to not paying capital recovery and variable operational charges. Here is the logic:

At Stage 1 in 1968, the time of ZESCO's creation, 75% of Zambia's population is dependent on primarily subsistence agriculture, yet mining, which provides employment to approx. 15% of the country's wage earners, contributes to 45–49% of Zambia's GDP (Park, 1968, pp. viii, 11). Light manufacturing contributes to approx. 9% of GDP (Park, 1968, p. vii). The Government of Zambia, owners of ZESCO, seek a strategy to develop Zambia's economy and improve quality of life. Hirschman's argument for unbalanced growth and focus on selected industries seems to have prevailed in the US Department of Commerce's analysis of Zambia in 1968 (Park, 1968, p. vii):

> Zambia's bright prospects for continuing economic development arise largely from the country's strong financial base, consisting principally of revenues derived from abundant mineral resources.

This thinking influenced the Government of Zambia's development strategy at Stage 2. It elected a strategy of unbalanced growth. It used its state-owned electricity utility ZESCO to subsidise productive sectors of the economy that would drive economic growth, with the hope that improved quality of life for all would follow. This explains why the World Bank found that ZESCO's tariffs in 1973 were low, and conditioned its loan on an increase of tariffs by 25% (IBRD and IDA, 1973, p. ii).

The past choice of subsidised tariffs informed and limited future decisions so that by Stage 3, Zambia offered tariffs that were 38% of the median sub-Saharan African electricity tariff (Trimble et al., 2016), whether to industry or to residential consumers.

By Stage 4 when ZESCO was negotiating tariffs with Nava Bharat, the premise of allocative efficiency for subsidised tariffs no longer held, because to support its customers' productivity, it needed to deliver reliable electricity. But low user tariffs had become an ossified norm (per Mahoney's prediction, 2000, pp. 513–14), and ZESCO had become unaccustomed to paying a capital recovery charge for new power generation assets, because the majority of its power supply came from assets that were constructed in the 1970s or earlier, and had therefore fully depreciated by the late 2000s. ZESCO had also become unaccustomed to paying a variable operating charge for fossil fuel, given that it was almost fully reliant on fuel-free hydropower. This is alluded to in its intention to return to cheaper, renewable energy sources from diesel generation (Zesco Ltd, 2017a, p. 7).[29] ZESCO's history of low user tariffs now prevented it from delivering on the promise of allocative efficiency because it was no longer able to finance the construction of new power generation and service delivery to its base of mining and manufacturing companies. Chapters 3 and 5 illustrated the costs to manufacturing firms of the power outages caused by this.

The critical juncture of the power outages of 2015 and 2016 resulted in two tariff increases in 2017. However, the large percentage increases on very small values meant that the increases were not substantial. They were not increased to be cost reflective (phone interview with an energy expert at a donor organisation on 27 February 2019). As the results of the manufacturing survey below show, the tariff increases did not affect manufacturing demand.

The critical juncture of a second wave of severe power outages, this time lasting 15 hours per day (Zesco, 2019), prompted the Energy Regulation Board to accept ZESCO's application to raise tariffs yet again on 1st January 2020 (see Fig. 5.7). Still, Haria and Ahmed (2020) note:

> The Energy Regulation Board commissioned a Cost of Service Study in 2016 with the findings on a cost-reflective electricity tariff due in 2017 (Energy Regulation Board of Zambia, 2016b). This remains outstanding, so those outside ZESCO and the ERB do not know whether the new tariffs are cost reflective. Still, their increase can only mean better system sustainability.

[29] There are three elements of a tariff that a power utility would pay an independent power project: the capital recovery charge, the fixed operations and maintenance charge and the variable operations and maintenance charge. The first includes the returns paid to equity and debt investors for their investment into the capital assets. The second covers salaries and other fixed operating costs. The third includes the cost of non-renewable fuels and other variable operating costs.

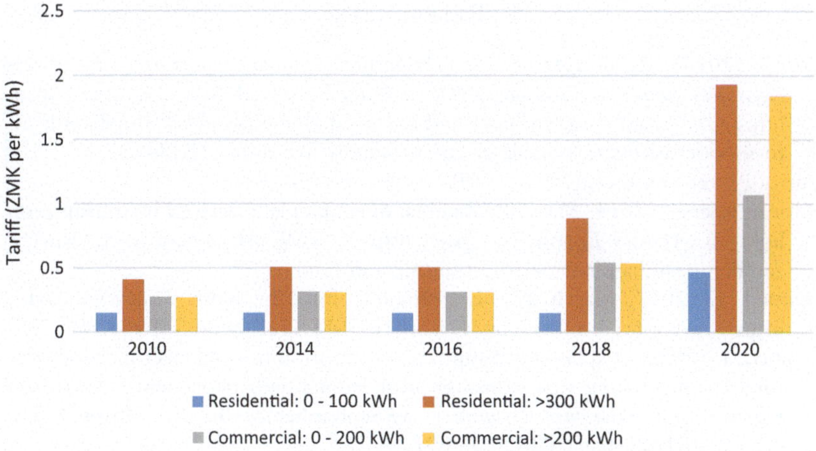

Fig. 5.7 Evolution of ZESCO tariffs, 2010–2020. (*Source*: Haria & Ahmed, 2020)

Resolving the issue of a system sustainable tariff would also ensure better creditworthiness for ZESCO vis-à-vis other power utilities and power projects on the SAPP, and ensure that ZESCO was better financially situated to pay for emergency power.

This chapter examined why Zambia's system of energy provision did not prevent the power outages of 2015, 2016 or even 2019. Warnings of climate change's effects on water availability for hydropower were available as early as the turn of the millennium so that as Zambia's system of energy provision planned to meet increasing demand, it should have taken into consideration a variety of climate futures. Even after the power outages of 2015 and 2016, even when the system recognises the need for technology or basin diversification, the system continues to look at the Batoka Gorge hydropower dam in the Zambezi River Basin whose potential was negatively assessed by Harrison and Whittington in 2002. Historical institutional path dependence explains this phenomenon. It also explains why a low tariff to end-consumers which had its beginnings in industrial policy persists in informing tariff policy that has hamstringed the system's ability to deliver reliable energy to Zambia's industrial base.

References

AfDB. (2013). *AfDB approves US $150-million loan to support access to reliable electricity supply in Zambia, AfDB website*. Retrieved May 25, 2020, from https://www.afdb.org/en/news-and-events/afdb-approves-us-150-million-loan-to-support-access-to-reliable-electricity-supply-in-zambia-12334

Africa Power Ventures. (2018). *SAPP market analysis*. Ebene.

African Energy. (2014, May 30). Zambia: Maamba plant delayed by transmission hold-up. *African Energy Newsletter*. https://www.africa-energy.com/article/zambia-maamba-plant-delayed-transmission-hold

Ahmed, I. (2017, March 27). Leapfrogging into the light. *Financial Times*. https://www.ft.com/content/41524878-108f-11e7-b030-768954394623

Ahmed, I. (2020). Explaining Rwanda's prioritisation of rural electrification over rural clean drinking water through institutional path dependency. *Structural Change and Economic Dynamics, 54*(September), 186–201. https://doi.org/10.1016/j.strueco.2020.05.001

Ahmed, I. (2021). *The political economy of energy mix in hydropower dependent developing nations–a case study of Zambia*. University College London. Available at: https://discovery.ucl.ac.uk/id/eprint/10121236/.

Ahmed, I., Parikh, P., Sianjase, G., & Coffman, D. (2020, December 4). The impact decades long dependence on hydropower in El Niño impact-prone Zambia is having on carbon emissions through diesel backup generation. *Environmental Research Letters, 15*(12) https://doi.org/10.1088/1748-9326/abb6a1

Al Jazeera. (2011, September 23). Opposition leader wins Zambia election. *Al Jazeera*. https://www.aljazeera.com/news/2011/9/23/opposition-leader-wins-zambia-election

Arup. (2017). *ASPIRE—A Sustainability Poverty and Infrastructure Routine for Evaluation*. https://sustainabilitytoolkit.arup.com/#/.

Beck, L., & Bernauer, T. (2011). How will combined changes in water demand and climate affect water availability in the Zambezi river basin? *Global Environmental Change, 21*(3), 1061–1072. https://doi.org/10.1016/j.gloenvcha.2011.04.001

Becker, M. C., et al. (2005). Applying organizational routines in understanding organizational change. *Industrial & Corporate Change, 14*(5), 775–791. https://doi.org/10.1093/icc/dth071

Beilfuss, R. (2012). *A risky climate for Southern African Hydro: Assessing hydrological risks and consequences for Zambezi River Basin Dams*. International Rivers. http://seine.oieau.fr/oieau.fr/sites/seine.oieau.fr.oieau.fr.eaudoc/files/32987.pdf

BEIS. (2019). *2019 Government greenhouse gas conversion factors for company reporting - Methodology paper for emission factors*.https://assets.publishing.service.gov.uk/government/uploads/system/uploads/attachment_data/file/829336/2019_Green-house-gas-reporting-methodology.pdf

Boyden, A., Kie, V., & Doolan, M. (2016). The environmental impacts of recycling portable lithium-ion batteries. *Procedia CIRP, 48*, 188–193. https://doi.org/10.1016/j.procir.2016.03.100

BRE Global. (2019). *CEEQUAL Version 6 technical manual.* BRE Global. https://www.ceequal.com/version-6/

Buragohain, T. (2012). Impact of solar energy in rural development in India. *International Journal of Environmental Science and Development, 3*(4), 334–338. http://www.ijesd.org/papers/242-B10021.pdf

Cecere, G., et al. (2014). Lock-in and path dependence: an evolutionary approach to eco-innovations. *Journal of Evolutionary Economics, 24*(5), 1037–1065. https://doi.org/10.1007/s00191-014-0381-5

Christensen, C. (1997). *The innovator's dilemma.* Harvard Business School Press.

Circular Energy Storage. (2019). *The lithium-ion battery end-of-life market-a baseline study, global battery alliance.* http://www3.weforum.org/docs/GBA_EOL_baseline_Circular_Energy_Storage.pdf

Coffman, D. (2013). Towards a New Jerusalem: The committee for regulating the excise, 1649–1653 Author (s): D'Maris Coffman Source: The, December 2013, Vol. 128, No. 535 (December Stable Url: http://www.jstor.com/stable/24473893 Towa. English Historical Review, *128*(535), 1418–1450. https://doi.org/10.1093/ehr/cet274

Coffman, D. et al. (2020). Ecological transition. In Cerniglia, F. and Saraceno, F. (eds.), *A European public investment outlook*, 161–174. https://doi.org/10.11647/OBP.0222

Colignon, R. A. (1997). *Power plays: Critical events in the institutionalization of the Tennessee valley authority.* State University of New York Press.

Conway, D., et al. (2015, September). Climate and southern Africa's water–energy–food nexus. *Nature Climate Change.* Nature Publishing Group, 5, 837–846. https://doi.org/10.1038/nclimate2735

Conway, D., et al. (2017). Hydropower plans in eastern and southern Africa increase risk of concurrent climate-related electricity supply disruption. *Nature Energy, 2*(12), 946–953. https://doi.org/10.1038/s41560-017-0037-4

Cornish, J.-J. (2019, August 24). South Africa scraps Russian nuclear plant plans. *Radio France International.* http://www.rfi.fr/en/africa/20190824-south-africa-cancel-russia-nuclear-plant-order-economy-ramaphosa

Dalberg Advisors, Lighting Global and World Bank. (2018). *Off-Grid Solar Market Trends Report 2018.* Washington, DC. https://doi.org/10.1017/CBO9781107415324.004

Davies, A., & Brady, T. (2000). Organisational capabilities and learning in complex product systems: Towards repeatable solutions. *Research Policy, 29*(7–8), 931–953. https://doi.org/10.1016/s0048-7333(00)00113-x

Deemer, B. R., et al. (2016). Greenhouse gas emissions from reservoir water surfaces: A new global synthesis. *Bioscience, 66*(11), 949–964. https://doi.org/10.1093/biosci/biw117

Diggers Editor. (2019, November 20). A history of debt, poor management of Zesco. *Diggers.* https://diggers.news/opinion/2019/11/20/a-history-of-debt-poor-management-at-zesco/

DNV GL. (2020). *Energy transition outlook 2020.* DNV GL.

Döll, P., et al. (2018). Risks for the global freshwater system at 1.5°C and 2°C global warming. *Environmental Research Letters, 13.* https://doi.org/10.1088/1748-9326/aab792

Dosi, G. (1982). Technological paradigms and technological trajectories. A suggested interpretation of the determinants and directions of technical change. *Research Policy, 11*(3), 147–162. https://doi.org/10.1016/0048-7333(82)90016-6

Dosi, G., Teece, D., & Al, E. (1992). Toward a theory of corporate coherence: Some preliminary remarks. In G. Dosi, R. Giannetti, & P. Toninelli (Eds.), *Technology and enterprise in a historical perspective.* Clarendon Press.

Energy Regulation Board. (2011). *Energy Sector Report 2010.*

Energy Regulation Board of Zambia. (2015). *Energy Regulation 2014.*

Energy Regulation Board of Zambia. (2016a). *Energy Sector Report 2015.* http://www.erb.org.zm/reports/EnergySectorReport2014.pdf

Energy Regulation Board of Zambia. (2016b). Request for expressions of interest: Electricity cost of service study and determination of economic cost-based tariffs. African Development Bank Group. https://www.afdb.org/fileadmin/uploads/afdb/Documents/Procurement/Project-related-Procurement/EOI_-_Zambia_-_Consulting_services_-_Electricity_Cost_of_Service_Study_and_Determination_of_Economic_Cost-based_Tariffs.pdf

Energy Regulation Board of Zambia. (2017a). *Energy Sector Report 2016.* https://doi.org/10.13140/RG.2.1.3893.7043

Energy Regulation Board of Zambia. (2017b). *Public Notice, April 2017.* http://www.erb.org.zm/downloads/zescoPCP2017FinalVersionDraft.pdf

Energy Regulation Board of Zambia. (2018). *Energy Sector Report 2017.* http://www.erb.org.zm/downloads/reports/esr2017.pdf

Energy Regulation Board of Zambia. (2019a). *Energy Sector Report 2018.* http://www.erb.org.zm/reports/esr2018.pdf

ESI-Africa. (2019, November). Zimbabwe includes pumped storage hydro in renewable energy strategy. *Hydro Review.* https://www.hydroreview.com/2019/11/18/zimbabwe-includes-pumped-storage-hydro-in-renewable-energy-strategy/#gref

Falchetta, G., et al. (2019). Hydropower dependency and climate change in sub-Saharan Africa : A nexus framework and evidence-based review. *Journal of Cleaner Production, 231*, 1399–1417. https://doi.org/10.1016/j.jclepro.2019.05.263

FAO. (2015). *FAO and El Niño, Food and Agriculture Oganization of the United Nations*. Retrieved March 28, 2020, from http://www.fao.org/el-nino/en/

FAO. (2018). *El Niño, Food and Agriculture Oganization of the United Nations*. Retrieved March 28, 2020, from http://www.fao.org/emergencies/crisis/elnino-lanina/intro/en/

FAO. (2020). *Aquastat database, Aquastat database*. Retrieved December 28, 2020, from http://www.fao.org/nr/water/aquastat/data/query/results.html

Ford, N. (2016, August). Gas fuels Tanzania, Mozambique growth. *The Africa Report*. https://www.theafricareport.com/939/gas-fuels-tanzania-mozambique-growth/

Funder, M., et al. (2018). The politics of climate change adaptation in development: Authority, resource control and state intervention in rural Zambia. *The Journal of Development Studies. Routledge, 54*(1), 30–46. https://doi.org/10.1080/00220388.2016.1277021

GHG Protocol. (2013). *Scope 3 calculation guidance, GHG protocol website*. Retrieved June 28, 2020, from https://ghgprotocol.org/scope-3-technical-calculation-guidance

Gibbs, J. (2019). *Planet of the humans*. Rumble Media and YouTube. https://www.youtube.com/watch?v=HQrZUj4-Xtk

Global Battery Alliance. (2019). *A vision for a sustainable battery value chain in 2030: Unlocking the full potential to power sustainable development and climate change mitigation*. http://www3.weforum.org/docs/WEF_A_Vision_for_a_Sustainable_Battery_Value_Chain_in_2030_Report.pdf

Gómez, D. R., et al. (2006). Chapter 2 stationary combustion. In A. Garg & T. Pulles (Eds.), *2006 IPCC guidelines for national greenhouse gas inventories Vol. 2 energy*. Institute for Global Environmental Strategies.

Government of Zambia. (2016). Zambia's intended nationally determined contribution to the 2015 agreement on climate change. UNFCCC NDC Registry. https://www4.unfccc.int/sites/ndcstaging/PublishedDocuments/ZambiaFirst/FINAL+ZAMBIA'S+INDC_1.pdf

Gunaratne, L., & LGA Consultants. (2002). *Overview of the electricity sector in relation to public private partnerships in Sri Lanka*. https://assets.publishing.service.gov.uk/media/57a08d4040f0b6497400174e/R8148-Ethioprep.pdf

Hall, P. A. (2010). Historical institutionalism in rationalist and sociological perspective. In J. Mahoney & K. Thelen (Eds.), *Explaining institutional change—Ambiguity, agency and power*. Cambridge University Press.

Hannan, M., & Freeman, J. (1989). *Organizational ecology*. Harvard University Press.

Hao, Z., et al. (2018, September). Quantifying the relationship between compound dry and hot events and El Niño—Southern Oscillation (ENSO) at the global scale. *Journal of Hydrology, 567*, 332–338. https://doi.org/10.1016/j.jhydrol.2018.10.022

Haria, S., & Ahmed, I. (2020). *No Increasing tariffs to prevent another electricity crisis in Zambia, International Growth Centre blog*. Retrieved May 7, 2020, from https://www.theigc.org/blog/increasing-tariffs-to-prevent-another-electricity-crisis-in-zambia/

Harmon, K. (2009, June 29). How does geothermal drilling trigger earthquakes? *Scientific American*. https://www.scientificamerican.com/article/geothermal-drilling-earthquakes/

Harrison, G. P., & Whittington, H. W. (2002). Susceptibility of the Batoka Gorge hydroelectric scheme to climate change. *Journal of Hydrology, 264*(1–4), 230–241. https://doi.org/10.1016/S0022-1694(02)00096-3

Henderson, R. M., & Clark, K. B. (1990). Architectural innovation: The reconfiguration of existing product technologies and the failure of established firms. Administrative Science Quarterly, *35*(Technology, Organizations and Innovation), 9–30. https://doi.org/10.2307/2393549

IBRD and IDA. (1970). *Appraisal of Kariba North Hydroelectric Project Zambia*. http://documents.worldbank.org/curated/en/917071468334477402/pdf/multi-page.pdf

IBRD and IDA. (1973). *Appraisal of Kafue Hydroelectric Project (Stage II) Zambia*. http://documents.worldbank.org/curated/en/824421468167973026/pdf/multi-page.pdf

IEA. (2020a). *Africa Energy Outlook 2019—Analysis Scenarios, Africa Energy Outlook 2019*, Paris. https://www.iea.org/reports/africa-energy-outlook-2019#energy-access; https://www.iea.org/reports/africa-energy-outlook-2019#africa-case

IEA. (2020b). *Global EV Outlook 2020, Global EV Outlook*. https://doi.org/10.1787/9789264302365-en

IMF. (2019). *Zambia: Gross domestic product, constant prices*. World Economic Outlook. https://www.imf.org/external/pubs/ft/weo/2019/02/weodata/weorept.aspx?sy=2017&ey=2024&scsm=1&ssd=1&sort=country&ds=.&br=1&pr1.x=49&pr1.y=11&c=754&s=NGDP_R&grp=0&a=

IPCC. (2000). *Summary for policymakers—Emissions scenarios*. Intergovernmental Panel on Climate Change. https://www.ipcc.ch/site/assets/uploads/2018/03/sres-en.pdf

IPCC. (2001). *Climate change 2001: Impacts, adaptation, and vulnerability*. https://www.ipcc.ch/site/assets/uploads/2018/03/WGII_TAR_full_report-2.pdf

IPCC. (2018). *Global warming of 1.5°C. An IPCC Special Report on the impacts of global warming of 1.5°C above pre-industrial levels and related global greenhouse gas emission pathways, in the context of strengthening the global response to the threat of climate change, Special Report*. https://www.ipcc.ch/site/assets/uploads/sites/2/2019/06/SR15_Full_Report_Low_Res.pdf

JICA. (2008). *Project formation study for power development planning in Zambia*. https://doi.org/10.1007/s00445-011-0503-y

Jury, M. (2002). Economic impacts of climate variability in South Africa and development of resource. *American Meteorological Society, 41*(1), 46–55. https://www.jstor.org/stable/26184931

Lazard. (2018). *Lazard's annual levelized cost of energy analysis—Version 12.0.* www.lazard.com

Levitt, B., & March, J. G. (1988). Organizational learning. *Annual Review of Sociology, 14*, 319–340. http://www.jstor.com/stable/2083321

Liebowitz, S. J., & Margolis, S. E. (1995). Path dependence, lock-in, and history. *Journal of Law, Economics, and Organization, 11*(1), 205–226. https://doi.org/10.2139/ssrn.1706450

Lighting Global, et al. (2020). *Off-grid solar market trends report 2020.* https://doi.org/10.1017/CBO9781107415324.004

Lighting Global and Vivid Economics. (2019). *Pay-as-you-go market attractiveness index 2019.* https://www.lightingglobal.org/wp-content/uploads/2019/06/PAYGo-MAI-2019-Report.pdf

Lofthouse, J., Simmons, R. T., & Yonk, R. M. (2015). Reliability of renewable energy: Geothermal. Logan. https://www.usu.edu/ipe/wp-content/uploads/2015/11/Reliability-Geothermal-Full-Report.pdf

Maamba Collieries. (2012). *Maamba collieries power generation project environmental and social impact assessment summary.*

Maamba Collieries. (2013). Maamba Collieries on schedule with power plant construction. *MCL News.* http://www.maambacoal.com/images/corporatemagazine/2013/Second2013.pdf

Maamba Collieries. (2015a, September). Grid upgrade works by ZESCO on course. *MCL News.* http://www.maambacoal.com/images/corporatemagazine/2015/Third2015.pdf

Maamba Collieries. (2015b). ZESCO starts power line upgrades to prepare for MCL power. *MCL News.* http://www.maambacoal.com/images/corporatemagazine/2015/Second2015.pdf

Mabumba, D. (2017). *Ministerial statement migration of electricity tariffs to cost-reflectivity.* National Assembly of Zambia. http://www.parliament.gov.zm/node/7020

Mahoney, J. (2000). Path dependence in historical sociology. *Theory and Society, 29*(4), 507–548. http://www.jstor.org/stable/3108585

Mar, L. E. (2009). *Carbon impact of proposed hydroelectric dams in Chilean Patagonia.* Massachusetts Institute of Technology. https://dspace.mit.edu/handle/1721.1/53068

Marchese, A. J., & Zimmerle, D. (2018, July 2). The US natural gas industry is leaking way more methane than previously thought. Here's why that matters. *The Conversation.* https://theconversation.com/the-us-natural-gas-industry-is-leaking-way-more-methane-than-previously-thought-heres-why-that-matters-98918

Matanzima, J. (2021). Thayer Scudder's Four Stage Framework, water resources dispossession and appropriation: The Kariba case. *International Journal of Water Resources Development, 00*(00), 1–24. https://doi.org/10.1080/07900627.2020.1866505

McKinsey. (2019). *Second-life EV batteries: The newest value pool in energy storage.* McKinsey & Company. https://www.mckinsey.com/industries/automotive-and-assembly/our-insights/second-life-ev-batteries-the-newest-value-pool-in-energy-storage

Mfula, C. (2016, May 10). Zambia to diversify generation mix as drought hits Hydropower. *Lusaka Times.* https://www.lusakatimes.com/2016/05/10/zambia-diversify-generation-mix-drought-hits-hydropower/

Ministry of Energy. (2018). *Strategic plan 2018–2021.* https://www.moe.gov.zm/wp-content/uploads/2019/09/Ministry-of-Energy-Strategic-Plan-2018-2021.pdf

Ministry of Energy and Water Development. (2008). *National energy policy.* https://doi.org/10.1017/CBO9781107415324.004

Ministry of Energy and Water Development. (2011). *Power system development master plan for Zambia.* https://www.moe.gov.zm/download/other-documents/Power-Systems-Development-Master-Plan.pdf

Ministry of National Development Planning. (2017). *Seventh National Development Plan 2017–2021.* http://www.lse.ac.uk/GranthamInstitute/wp-content/uploads/2018/09/8721.pdf

National Grid. (2015). *Electricity capacity report.* https://doi.org/10.1016/j.matchemphys.2007.04.026

Ncube, M., & Lufumpa, C. L. (2017). *Infrastructure in Africa—Lessons for future development.* Policy Press.

Nelson, R., & Winter, S. (1982). *An evolutionary theory of economic change.* Harvard University Press.

North, D. C. (1990). Institutions, institutional change and economic performance—Political economy of institutions and decisions. Kindle. University of Cambridge Press.

Ocko, I. B., & Hamburg, S. P. (2019). Climate impacts of hydropower: Enormous differences among facilities and over time. *Environmental Science and Technology, 53*(23), 14070–14082. https://doi.org/10.1021/acs.est.9b05083

Olivero, J., et al. (2017). Recent loss of closed forests is associated with Ebola virus disease outbreaks. *Nature Scientific Reports, 7*(1), 1–9. https://doi.org/10.1038/s41598-017-14727-9

Ovo Energy. (2018). *Nuclear power explained, Ovoenergy.com.* Retrieved January 1, 2021, from https://www.ovoenergy.com/guides/energy-sources/nuclear-energy.html#:~:text=The potential power locked inside,create aridiculous24%2C000%2C000kWh.

Park, S. F. (1968). *Basic data on the economy of Zambia.* https://play.google.com/books/reader?id=K_ZhUrU5lfoC&hl=en_GB&pg=GBS.PA1

Patil, R. (2017, July 28). Nava Bharat Ventures up 2%, achieves provisional acceptance of 300 MW power plant. *Money Control.* https://www.moneycontrol.com/news/business/stocks/nava-bharat-ventures-up-2-achieves-provisional-acceptance-of-300-mw-power-plant-2338365.html

Pierson, P., & Skocpol, T. (2002). Historical institutionalism in contemporary political science. In Katznelson, I. and Milner, H. (eds.), *Political Science—State of the Discipline*, pp. 693–721. American Political Science Association and WW Norton & Company.

Power Technology. (2013, October 6). AfDB approves $150m loan for 300 MW Maamba power project and coal mine in Zambia. *Power Technology.* Retrieved from https://www.power-technology.com/uncategorised/newsafdb-approves-150m-loan-for-300mw-maamba-power-project-and-coal-mine-in-zambia/

Pravalie, R., & Bandoc, G. (2018). Nuclear energy: Between global electricity demand, worldwide decarbonisation imperativeness, and planetary environmental implications. *Journal of Environmental Management, 209*, 81–92. https://doi.org/10.1016/j.jenvman.2017.12.043

Resnick, D., & Thurlow, J. (2014). *The political economy of Zambia's recovery: Structural change without transformation?*http://ebrary.ifpri.org/utils/getfile/collection/p15738coll2/id/128012/filename/128223.pdf

Richter, A. (2020, May 13). Developer Kalahari GeoEnergy Ltd. has secured a convertible loan of USD 3.2 m for a up to three slim wells to be drilled for the company's Bweengwa geothermal power project in Zambia, Africa. *Think GeoEnergy.* Retrieved from https://www.thinkgeoenergy.com/kalahari-geoenergy-secures-3-2m-loan-funding-for-geothermal-project-in-zambia/

Sabatier, P. A. (1998). The advocacy coalition framework: Revisions and relevance for Europe. *Journal of European Public Policy, 5*(1), 98–130. https://doi.org/10.1080/13501768880000051

SADC. (2008). *Integrated water resources management strategy and implementation plan for the Zambezi River Basin.* http://www.zambezicommission.org/sites/default/files/clusters_pdfs/Zambezi%20River_Basin_IWRM_Strategy_ZAMSTRAT.pdf.

Sandwell, P., et al. (2016). Analysis of energy access and impact of modern energy sources in unelectrified villages in Uttar Pradesh. *Energy for Sustainable Development. International Energy Initiative, 35*, 67–79. https://doi.org/10.1016/j.esd.2016.09.002

Scherer, L., & Pfister, S. (2016). Hydropower's biogenic carbon footprint. *PLoS ONE, 11*(9), 1–12. https://doi.org/10.1371/journal.pone.0161947

Scudder, T., 1993. Pipe Dreams: Can the Zambezi River supply the region's water needs?. Cultural Survival Quarterly, [online] (June). Available at: [Accessed 23 June 2021].

Scudder, T., & Colson, E. (1982). From welfare to development: A conceptual framework for the analysis of dislocated people. In A. Hansen & A. Oliver-

Smith (Eds.), *Involuntary migration and resettlement* (pp. 267–288). Westview Press.

Serpell, N. (2019). *The old drift* (1st ed.). Hogarth.

Sinclair-Desgagné, B., & Soubeyran, A. (2000). *A Theory of Routines as Mindsavers*, 52. Montréal. https://www.researchgate.net/publication/4816610_A_Theory_of_Routines_as_Mindsavers

Sinkamba, K. (2011, September 2). Zambia approves power purchase agreement between ZESCO and Maamba Collieries. *SteelGuru*. https://steelguru.com/steel/zambia-approves-power-purchase-agreement-between-zesco-and-maamba-collieries/223295

Spalding-Fecher, R. (2012). *Powering Southern Africa—The future of hydro in the Zambezi River Basin, Climate & Development Knowledge Network*. Retrieved June 6, 2020, from https://cdkn.org/2012/02/powering-southern-africa-the-future-of-hydro-in-the-zambezi-river-basin/?loclang=en_gb

Spalding-Fecher, R., et al. (2016). The vulnerability of hydropower production in the Zambezi River Basin to the impacts of climate change and irrigation development. *Mitigation and Adaptation Strategies for Global Change, 21*(5), 721–742. https://doi.org/10.1007/s11027-014-9619-7

Stinchcombe, A. L. (1968). *Constructing social theories*. University of Chicago Press.

Takouleu, J. M. (2020, May 25). REPP finances drilling of 3 geothermal wells on Bweengwa site. *Afrik21*. https://www.afrik21.africa/en/zambia-repp-finances-drilling-of-3-geothermal-wells-on-bweengwa-site/

Teljeur, E., Chetty, M., & Hendriksz, M. (2017). Africa's prospects for infrastructure development and regional integration: Energy sector. In M. Ncube & C. L. Lufumpa (Eds.), *Infrastructure in Africa—Lessons for Future Development* (1st ed., pp. 185–256). Policy Press.

Tembo, B., & Merven, B. (2013). Policy options for the sustainable development of Zambia's electricity sector. *Journal of Energy in Southern Africa, 24*(2), 16–27. https://doi.org/10.17159/2413-3051/2013/v24i2a3126

Trimble, C. et al. (2016). *Financial viability of electricity sectors in sub-Saharan Africa—Quasi-fiscal deficits and hidden costs, energy and extractives global practice group*, 7788. https://openknowledge.worldbank.org/handle/10986/24869

Tripsas, M., & Gavetti, G. (2000). Capabilities, cognition, and inertia: Evidence from digital imaging. *Strategic Management Journal, 21*(Special), 1147–1161. https://doi.org/10.1002/1097-0266(200010/11)21:10/11<1147::aid-smj128>3.0.co;2-r

Tushman, M. L., & Anderson, P. (1986). Technological discontinuities and organizational environments. *Administrative Science Quarterly, 31*(3), 439–465. https://doi.org/10.2307/2392832

United Nations. (2017). *Household size and composition around the world, Economic & Social Affairs*. https://doi.org/10.3390/atmos6091362

US EPA. (2020a). *Understanding Global Warming Potentials, US Environmental Protection Agency website*. Retrieved June 6, 2020, from https://www.epa.gov/ghgemissions/understanding-global-warming-potentials#:~:text=Methane (CH4) is estimated,uses a different value.).&text=Nitrous Oxide (N2O,for a 100-year timescale

US EPA. (2020b). *Why is acid rain harmful? US Environmental Protection Agency website*. Retrieved August 10, 2020, from https://www3.epa.gov/acidrain/education/site_students/whyharmful.html#:~:text=Acid Rain Harms Forests,trees to take up water

USAID, Tony Blair Institute for Global Change and Power Africa. (2019). *West Africa Power Trade Outlook*. https://institute.global/sites/default/files/inline-files/ENGLISHWestAfricaPowerTradeUKWHOLELOW_1.pdf

Vidal, J. (2016, May 30). El Niño is over—But it leaves nearly 100 million people short of food. *The Guardian*. https://www.theguardian.com/global-development/2016/may/30/el-nino-is-over-but-it-leaves-nearly-100-million-people-short-of-food

Vivid Economics and Oeko-Institut. (2020). *Closing the loop on energy access: Analytical study*. Vivid Economics.

Wang, G., et al. (2017). Continued increase of extreme El Niño frequency long after 1.5°C warming stabilization. *Nature Climate Change, 7*(8), 568–572. https://doi.org/10.1038/NCLIMATE3351

Waylen, P., Annear, C., & Bunting, E. (2019). Interannual hydroclimatic variability of the lake. *Water, 11*(9), 1–14. https://doi.org/10.3390/w11091801

Willing, N. (2020, March). Chinese 5G rollout lifts lithium battery demand. *Argus Media*. https://www.argusmedia.com/en/news/2088490-chinese-5g-rollout-lifts-lithium-battery-demand

World Bank. (1999). Project appraisal document on a proposed loan and GEF grant in the amount of IBRD loan US$ 30 million and GEF Grant SDR 7.2 million to Argentine Republic for a Renewable Energy in the Rural Market Project. http://documents.worldbank.org/curated/en/271841468768697115/pdf/multi-page.pdf

World Bank. (2017a) *Combined project information documents—Electricity Services Access Project P162760*. PIDISDSA21985.

World Bank. (2017b). *Investor brief*. DC. http://pubdocs.worldbank.org/en/284471507050951684/investor-brief-worldbank-2017.pdf

World Bank. (2018). Zambia precipitation 1991–2015. Climate Change Knowledge Portal, World Bank Group. http://sdwebx.worldbank.org/climateportal/index.cfm?page=downscaled_data_download&menu=historical

World Bank. (2019). *World development indicators: Zambia*. World Bank. https://data.worldbank.org/country/zambia

World Bank (2020a) *GDP (constant LCU)*. https://data.worldbank.org/indicator/NY.GDP.MKTP.KN

World Bank. (2020b). *Projects in Bolivia, World Bank.* Retrieved October 17, 2020, from https://projects.worldbank.org/en/projects-operations/projects-list?lang=en&searchTerm=&countrycode_exact=BO

World Bank. (2020c). *Zambia world development indicators.*

Yamba, F. D., Walimwipi, H., & Mzezewa, C. (2011). Climate change / variability implications on hydroelectricity generation in the Zambezi River Basin. *Mitigation and Adaptation Strategies for Global Change, 16*(6), 617–628. https://doi.org/10.1007/s11027-011-9283-0

Zambeef Products. (2017). *Annual Report 2016.*

Zambia Bata. (2016). *Annual financial statements for the year ended 31 December 2015.*

Zambia Bata. (2017). *Annual Report 2016.*

Zambia Bata. (2018). *Abridged Report 2017.*

Zambia Meteorological Department. (2018). *Raw precipitation data of 39 sites in Zambia, 1991–2018.*

Zambia Sugar. (2017). *2016 Annual Report.* https://africanfinancials.com/document/zm-zmsg-2016-ar-00/

Zesco. (2019). *Countrywide load-shedding schedule October 2019 revised.* http://www.zesco.co.zm/customerCare/loadSheddingSchedule

Zesco Ltd. (2017a). *2016 Integrated Report.*

Zesco Ltd. (2017b). *Zesco's planned additional energy generation.* Zesco Ltd..

Zesco Ltd. (2018). 2017 Integrated report.

ZICTA and CSO. (2018). *2018 National survey on access and usage of information and communication technologies by households and individuals—A demand side assessment of access and usage of ICTs in Zambia.* https://www.zicta.zm/Downloads/publications/2018ICTSurvey_PreliminaryReport.pdf

CHAPTER 6

Recommendations

Chapter 2 reviewed the historic role that industrialisation has played in the economic development of nations. It observed that manufacturing grew in Zambia following investments into power generation until 1991, but that its growth had minimal impact on structural changes in employment shifts away from agriculture and in fact GDP per capita was lower by 1991 than it was at the time of independence in 1964, with average life expectancy at about the same level. Since the 2000s, manufacturing has been growing again, with higher employment and GDP returns per kWh of energy consumed than Zambia's largest consumer of energy, mining. Chapter 4 showed the ways in which power outages harm Zambian manufacturing firms, and investigated manufacturing firms' interventions to lessen the costs of power outages and investigated the efficacy of their interventions. Chapter 5 assessed distribution and transmission losses, diligenced forecast demand and considered viable alternatives to hydropower. It explained the reasons why Zambia's system of energy provision did not prevent these power outages. This chapter will offer recommendations on how to overcome Zambia's power outages in the future in the hope that they offer a more equitable development pathway.

I presented some recommendations earlier in written format (Ahmed et al., 2019); in a closed-door presentation at the Ministry of National Development Planning to key representatives from Zambia's Ministry of Finance, Ministry of Energy, Energy Regulation Board, ZESCO and Central Statistical Office (4 September 2019); in the keynote presentation

I. Ahmed, *The Political Economy of Hydropower Dependant Nations*,
https://doi.org/10.1007/978-3-030-71266-2_6

of a public forum hosted by the International Growth Centre (Ahmed, 2019) and on prime-time TV news (ZNBC, 2019). Some of those recommendations have been updated here because the Energy Regulation Board subsequently accepted ZESCO's application for a further tariff escalation on 1 January 2020. To systematically approach the problems that can be solved, this chapter will offer solutions to address supply, and then offer solutions to dampen demand.

6.1 Supply-Side Solutions

ZESCO and the Energy Regulation Board should investigate ZESCO's increasing sources of distribution and transmission losses and work to address these before commissioning extra power supply. Additionally, before investing in or committing to purchase additional power supply, ZESCO should work to improve existing hydropower capacity through rehabilitation, refurbishment and renovation.

Delivering a more reliable energy supply while achieving sustainability is an example of a wicked problem: because of complex interdependencies, the effort to solve one aspect of a wicked problem may reveal or create other problems. Addressing Zambia's issue of power supply reliability by recommending full reliance going forward on coal-fired power plants adds to the accumulation of greenhouse gases in the atmosphere that have exacerbated the El Niño effect, which is causing ZESCO's existing hydropower infrastructure to become less effective. It would also acidify rain, which would damage vegetation and thus limit transpiration, interrupting a necessary process of putting water vapour into the atmosphere for rainfall to occur, as well as harm human health. To overcome the problem, we categorised possible sources of energy by a number of essential factors to consider: availability, dependability, climate impact, impact on health and the environment and financial costs in Table 5.6. The analysis shows that a more balanced portfolio is required, and that ZESCO will need to supply approximately 19,000 GWh of power by 2030. Realistically, gas-fired power imported from Zambia's SAPP neighbours Tanzania and Mozambique seem to be the most viable short-term options if ZESCO is able to charge its customers the costs of importing energy. For the medium term, ZESCO should look into the feasibility of pairing hydroelectricity reservoirs with solar-powered pumps for recharging. It should also follow-developments in the costs of battery storage that could make solar PV a viable baseload source of energy. ZESCO should also spear-head expanding the SAPP beyond southern Africa so that weather patterns affecting one river basin can be offset by alternative weather patterns in other parts of Africa. Diversifying across time zones will also make solar without storage more viable.

ZESCO will need to charge a cost-recovery tariff. The Government of Zambia does not have the resources to subsidise ZESCO and ZESCO does not have the resources to subsidise its customers and additionally increase grid electricity access. It should not have taken more than four years for the Energy Regulation Board's 2016 commission of the Cost of Service Study to be completed. It is likely that when it is completed, it will find that the cost-recovery tariff for ZESCO exceeds the average tariff ZESCO charges since the Energy Regulation Board approved the revised set of tariffs on 1st January 2020. Raising the tariff to the cost-recovery level will allow ZESCO to procure the additional energy it needs to serve its customers reliably, whether domestically or from the power pool, and achieve system sustainability. The manufacturing survey, which monetised the minimum impact of power outages for diesel generators in 2016 as USD 0.25/kWh, shows that low tariffs to the manufacturing sector are not achieving the goal of allocative efficiency. The same goes for the mining sector which also employs diesel generators for backup energy. With access to grid electricity only available to a third of the population, subsidised tariffs are inherently regressive against those without grid-electricity since they financially disable ZESCO from expanding its distribution network. Until all poor and rural communities in Zambia have access to grid-electricity, subsidised tariffs do not make sense from a distributional equity perspective. Nor should cost-recovery tariffs be restricted to households. Parliament should ensure that the Energy Regulation Board is empowered to set cost-recovery tariffs for the mining sector.

For ZESCO to charge yet higher tariffs that will enable it to achieve system sustainability, it will need to build trust with its clients. A quarter of manufacturing firms reported not receiving accurate load shedding notifications during the period April–August 2018. ZESCO must ensure therefore to the extent that it can that its communications to its clients are correct. To facilitate accurate communication about load shedding schedules, perhaps it will need to take more conservative estimates of the reservoir levels of its hydropower dams and their capacity utilisation rates. To dispel notions that other sectors subsidise mining, ZESCO should also make public on its website the tariffs that its mining clients and the Copperbelt Energy Corporation receive; the Copperbelt Energy Corporation in turn should disclose what tariffs it charges to its clients.

Given that they self-generate electricity more, food and beverages and export-facing firms within the manufacturing sector incur greater financial costs in running their businesses normally. They also emit more greenhouse gases with that self-generation (Ahmed et al., 2020). ZESCO should consider offering these firms a premium service solution, and link these firms' energy supply to nuclear power in South Africa. Alternatively, to offset the prioritisation of steady Zambian coal-fired power, ZESCO could require manufacturing firms to install solar panels (notably absent on manufacturing firms' warehouse and factory rooftops), or to finance renewable energy interventions.

6.2 Solutions That Can Dampen Demand for Electricity or Improve Self-sufficiency

The Ministry of Finance's tax waiver of generators seems to have been responsible for increased investment in generators. These help sectors of the economy limit the damages caused by power outages. Until ZESCO can ensure that the food and beverage manufacturing subsector and that export-facing manufacturing firms have uninterrupted power supply, the Ministry of Finance should retain these waivers to ensure smoother business operations. Similarly, other interventions should be introduced to encourage firms to generate their own energy, especially if it leads to a net reduction in harm the environment (such as some forms of waste-to-energy which are less greenhouse gas emitting and whose disposal processes would otherwise be greatly harmful), or a reduction in emissions (such as solar panels on rooftops). Feed-in-tariffs seem to have been a popular and successful mechanism for encouraging such investment.[1]

The differential between ZESCO's new peak and off-peak tariffs continues to be minimal. The peak tariff should be increased further to dampen demand during peak hours, so that ZESCO does not have to procure additional power capacity that will only be used a few hours of the day. Similarly, to reduce wastage, ZESCO should impose a surcharge on

[1] The South African utility Eskom does not seem to cater a feed-in-tariff for households (Strydom & Morar, 2018). However, nine municipalities in the Western Cape are piloting feed-in-tariffs for households. This is perhaps responsible for the Western Cape having the most solar rooftop installations (301) in South Africa (Strydom & Morar, 2018). The Eastern Cape has one municipality with an approved feed-in-tariff and has the second most installations (154) (Strydom & Morar, 2018). In the provinces without feed-in-tariffs, no PV installations have been recorded (Strydom & Morar, 2018).

reactive power, the power that flows back from a destination towards the grid in an alternating current scenario. Increasing tariffs beyond cost-recovery will additionally have the effect of incentivising greater energy efficiency, which the household survey has shown has been lacking.

References

Ahmed, I. (2019). Keynote presentation. In IGC (Ed.), *Nurturing the manufacturing sector: Ensuring strong roots for Zambian growth.* International Growth Centre.

Ahmed, I., Baddeley, M., Coffman, D., Meikle, J., Oseni, M., & Sianjase, G. (2019). *The cost of power outages to Zambia's manufacturing sector.* https://www.theigc.org/wp-content/uploads/2019/08/Ahmed-et-al-2019-policy-brief-41408.pdf. https://doi.org/F-41408-ZMB-1

Ahmed, I., Parikh, P., Sianjase, G., & Coffman, D. (2020, December 4). The impact decades long dependence on hydropower in El Niño impact-prone Zambia is having on carbon emissions through diesel backup generation. *Environmental Research Letters, 15*(12) https://doi.org/10.1088/1748-9326/abb6a1

Strydom, A., & Morar, G. (2018). *Rooftop solar: While municipalities focus on feed-in tariffs, a much larger problem persists, Sustainability Institute.* Retrieved March 17, 2020, from https://www.sustainabilityinstitute.net/si-news/5503-rooftop-solar-while-municipalities-focus-on-feed-in-tariffs-a-much-larger-problem-persists

ZNBC. (2019). *Nurturing the manufacturing sector: Ensuring strong roots for Zambian growth.* ZNBC. https://www.youtube.com/watch?v=4VTlhgi713w&t=11s

CHAPTER 7

Conclusions

Hydropower dependence makes countries vulnerable to climate change. As the El Niño effect increases in frequency and intensity as global temperatures rise, it causes an increase in droughts in Southern Africa, north South America and Southeast Asia. Whereas systems of provision have in cases like Zambia's taken for granted that hydropower is a dispatchable source of power, they should no longer take it for granted in climate-vulnerable geographies.

Donor-funded energy from the 1970s primarily targeted energy for mineral extraction in Zambia to the extent that power generation constructed in Zambia was constructed for the purpose of consumption in Zambia (as mentioned it Chap. 2, it was also intended for sanctioned Southern Rhodesia which could not directly access World Bank financing), perpetuating a colonial trend of meeting Western market needs. It was premised on the argument that unbalanced growth would pay for so-called social infrastructure and benefit the rest of the population in the long term. Metrics evaluating Zambia's development since the 1970s do not support this investment thesis. Most Zambians still cannot access the power supply that the World Bank financed in the 1960s and 1970s, which still accounts for the majority of Zambia's installed power capacity. Zambia's real GDP has grown 450%, but real GDP per capita, for all its faults as a metric, has grown only 8% since 1964. Income inequality has grown. Life expectancy has finally increased, but only after reaching a

I. Ahmed, *The Political Economy of Hydropower Dependant Nations*,
https://doi.org/10.1007/978-3-030-71266-2_7

nadir in the mid-1990s; it is not directly attributable to the World Bank's investments.

Path dependence within Zambia's energy system of provision resulted in ZESCO being insufficiently prepared for the climate events of 2015, 2016 and 2019, which rendered low hydropower reservoir levels. Further investment in hydropower failed to build its climate resilience. Retaining low user tariffs precluded Zambia's system of provision from sufficiently financing the construction of additional power generation capacity or sourcing it from abroad to avert power outages. Low user tariffs therefore harmed the very users Zambia's system of energy provision was subsidising.

The harm to manufacturing firms included increased staff costs, damage to their equipment, damage to their reputation, increased security costs, as well as damage to inventory. To mitigate against the harm of power outages, mining companies, manufacturing firms (at the additional cost of USD 0.25/kWh) and households, and even ZESCO itself have resorted to generating energy by burning diesel. This increases Zambia's emissions by up to 27,000 tonnes of carbon dioxide in months of severe outages. This incrementally contributes to a rise in global temperatures which in turn increases the likelihood of El Niño droughts that further undermine Zambia's existing hydropower infrastructure, and also undermines the government's effort to reduce the nation's emissions per its Paris Agreement Nationally Determined Contributions commitment.

ZESCO, the Energy Regulation Board, Ministry of Energy, donor agencies and financial institutions should broaden their criteria for selecting and supporting new energy sources. These should include the lifecycle environmental, health and climate impacts, availability, dependability and climate resilience and financial lifecycle costs in a 'least-regret' framework. This broader framework for energy investment appraisal would help solve Zambia's climate vulnerability.

With the right governance structures in place to prevent hold-ups and renegotiations, Zambia's best chances of achieving least-regret energy are likely through the import of emission-free nuclear power from South Africa, to the extent that additional nuclear generation capacity can be procured by Eskom through good governance in order to increase the likelihood of safe custodianship. In case such procurement through good governance is not attained, the next least-regret solution for is an intermediate one: imported natural gas fired-power from Tanzania and Mozambique. Although methane leakages through its value chain make natural gas' adverse climate impact comparable to that of coal, its adverse impacts on human health and ecology are not nearly as bad. Natural gas would have to be the intermediate solution for meeting baseload demand throughout the day as well as dispatchable peak demand until low-cost (financially and environmentally) batteries become available and make intermittent solar PV and wind energy viable as dispatchable energy sources. If pairing solar PV and wind with Zambia's existing hydropower infrastructure proves feasible, solar and wind could be brought into play even earlier to contribute to a low-carbon, climate-resilient source of dispatchable energy through pumped-storage.

The case of reduced hydropower capacity in Zambia is unfortunately not a unique one. Up to a fifth of the global population live in nations heavily dependent on hydropower that are vulnerable to drought due to increases in average global temperatures. As droughts and El Niño events increase in frequency and intensity across the world, the case study of and recommendations for Zambia have significant implications for the formulation of low-carbon energy transition policy in these other countries.

Appendix A: Hydropower-Dependent Countries That Like Zambia Experienced El Niño Droughts

Country/ territory	*Column A Conditions applied to World Bank Data (2020s): Not 'High Income' AND >25% electricity production comes from hydropower sources = 1; >25% and High Income = 0*	*Column B Source that says that countries that score 1 in column A experiences El Niño droughts*	*Column C If Column A = 1 AND affected by El Niño droughts, evidence of similar political economy dynamics viz. early World Bank focus on hydropower?*	*Column D Population (World Bank, 2020s) if not >25% hydropower AND are not High Income AND experience El Nino droughts*
Total countries that meet condition	45	32	17	1,357,770,729
Angola	1			
Albania	1			
Argentina	1	Rojas et al. (2014)		43,590,368
Armenia	1			
Austria	0			
Bosnia and Herzegovina	1			

(*continued*)

I. Ahmed, *The Political Economy of Hydropower Dependant Nations*, https://doi.org/10.1007/978-3-030-71266-2

(continued)

Bolivia	1	FAO (2016)	The World Bank's first investments into Bolivia were into hydropower plants (IDA, 1964; World Bank, 2020d)	11,031,813
Brazil	1	Rojas et al. (2014)	The World Bank's second investment into Brazil in 1950 was in hydropower followed by investment in another hydropower station in 1953 to meet industrial demand (IBRD, 1950c, 1953a; World Bank, 2020e).	206,163,058
Canada	0			
Cambodia	1	Rojas et al. (2014)		15,766,293
Cameroon	1	European Commission (2015)		23,926,539
Chile	0	Cai et al. (2020)		18,209,068
Congo, Dem. Rep.	1	WFP (2016) and Isango (2017)		78,789,127
Congo, Rep.	1			

(continued)

(continued)

Colombia	1	Rojas et al. (2014)	The World Bank's first investments in Colombia were in hydropower, starting from its second investment, to meet industrial demand (IBRD, 1950a, 1950b, 1951a; World Bank, 2020f). The World Bank's choice of investment in power over 'social' investment in water has been discussed in greater length in footnote 12.	48,171,392
Costa Rica	1	Costa Rica Guide (2020)	The World Bank's fourth investment in Costa Rica was in hydropower (IBRD, 1961b; World Bank, 2020g).	4,899,345
Croatia	0			
Ecuador	1	Rojas et al. (2014)	The World Bank's second investment in Ecuador was in 3 MW thermal and 14.5 MW hydropower plants (IBRD, 1956a; World Bank, 2020h) to meet industrial and residential demand.	16,491,115

(*continued*)

(continued)

Ethiopia	1	Rojas et al. (2014)	The World Bank's eighth investment in Ethiopia was in power to meet expanding industrial demand (IBRD, 1961c; World Bank, 2020i).	103,603,501
Gabon	1			
Georgia	1			
Ghana	1	Rojas et al. (2014)	The World Bank's first investments in Ghana were into the Volta Power Project (World Bank, 2020j) for the purpose of 'the aluminum smelter, various mining companies and the Electricity Division which would distribute power to the general consumers in southern Ghana' (IBRD, 1961a, p. 2).	28,481,946

(*continued*)

(continued)

Guatemala	1	Lakhani (2019)	The World Bank's second and third investments in Guatemala were in hydropower, with the Bank noting that industrial electricity sales accounted for 42% of electricity in 1966 (IBRD, 1967; IBRD and IDA, 1968, p. 4; World Bank, 2020k)	15,827,690

(*continued*)

(continued)

Honduras	1	Moloney (2019)	The World Bank's third investment in Honduras was for an interim power project consisting of the installation of additional diesel capacity and the preliminary costs for survey work for the Rio Lindo hydroelectric scheme. The Bank's fourth investment in Honduras was in the first stage of the Lake Yojoa-Rio Lindo hydroelectric power project (IBRD, 1959, 1960b; World Bank, 2020l). The target customers were mines, residential communities, a railroad company and a cement factory (IBRD, 1960b, p. 2).	9,270,795
Iceland	0			

(*continued*)

(continued)

Kenya	1	Rojas et al. (2014)	The World Bank's first investment into energy was in hydropower (IBRD and IDA, 1971, p. ii). Coming as it did in 1971, it lagged many other Bank investments in Kenya. This was because electricity demand was driven by industrial growth which was slow prior to 1971, but 1971 represented an inflection point in Kenya's industrialisation (IBRD and IDA, 1971, p. 3). No fossil fuels had been discovered in Kenya as of the time of investment, while hydropower was assessed as abundant (IBRD and IDA, 1971, p. 2).	49,051,686
Kyrgyz Republic	1			
Latvia	0			
North Macedonia	1			
Montenegro	1			
Mozambique	1	Rojas et al. (2014)		27,829,942

(*continued*)

(continued)

Myanmar	1	Sutton et al. (2019)		53,045,226
Namibia	1	Rojas et al. (2014)		2,358,041
Norway	0			
Nepal	1			
Nauru				
New Zealand	0			
Oman				
Pakistan	1	Rojas et al. (2014)		203,627,284
Panama	0	Fountain (2019)	The World Bank's fifth investment in Panama was in hydropower (World Bank, 2020m). Unlike other early power investments, this was for rural electrification to stem depopulation of rural areas (IBRD, 1962).	4,037,078
Peru	1	Rojas et al. (2014)	The World Bank's 12th investment, its first in power, was in hydropower (World Bank, 2020n). The power was intended to meet demand in Lima (IBRD, 1960a)	30,926,032
Korea, Dem. People's Rep.	1	FAO (2015a)		25,307,744
Portugal				
Paraguay	1	Rojas et al. (2014)		6,777,872
Romania	1			
Slovenia	0			
Sri Lanka	1			

(*continued*)

(continued)

Sudan	1	Rojas et al. (2014)	The World Bank's fifth investment into Sudan was in power generation and transmission (World Bank, 2020o). Two thirds of the installed capacity was in hydropower and a third was in a thermal power plant (IBRD and IDA, 1961).	39,847,440
Suriname	1	FAO (2016)		564,888
Sweden	0			
Switzerland	0			
Tajikistan	1			
Tanzania	1	Rojas et al. (2014)		53,050,790
Togo	1			
Turkey	1	Rojas et al. (2014)	The World Bank's fourth investment in Turkey was in a multi-purpose dam which helped control floods, expanded irrigation and provided power for expanding industries (IBRD, 1952b; World Bank, 2020p).	79,821,724

(*continued*)

(continued)

Uruguay	0	Rojas et al. (2014)	The World Bank's third investment in Uruguay was in hydropower, but this was also the Bank's third investment in power—the first and its extension being in thermal energy (IBRD, 1956b; World Bank, 2020q).	3,424,132
Venezuela, RB	1	Rojas et al. (2014)	The World Bank's first investment in Venezuela was in hydropower (World Bank, 2020r).	29,846,179
Vietnam	1	FAO (2015a)		93,638,724
Zambia	1	Rojas et al. (2014)	See chapter 3.	16,363,507
Zimbabwe	1	Rojas et al. (2014)	The World Bank's first investments in what was then a British colony, Rhodesia, also referred to as 'the Colony', were in hydropower in the Kariba hydropower project (IBRD, 1952a, 1956c; IBRD and IDA, 1964).	14,030,390

Appendix B: Last Iteration of Manufacturing Questionnaire Used, July 2018

I. Ahmed, *The Political Economy of Hydropower Dependant Nations*, https://doi.org/10.1007/978-3-030-71266-2

POWER OUTAGES AND MANUFACTURING FIRMS' PERFORMANCE IN ZAMBIA

Important Instruction: The items in this **questionnaire** ask about the performance of electricity supply, how it affects firms' operation, and your establishment's dispositions towards improved electricity service. Answers will be received in complete confidence. The researcher is only interested in the data obtained, not in who you are. Please be frank and respond as honestly as possible as this research is completely non-judgmental. Please note that all responses are coded and general trends reported so that no individual is identified in write ups of the survey.

Background firm information to be collected pre-interview

Name of enumerator(s):

Name of firm:

Address(es) (district(s), province(s)):

MFEZ, Light or Heavy Industrial Zone or none of the above?

Size of firm by employees:

Maximum power load consumed by firm, kVA:

What does the firm produce?:

Industry:

Date of survey:

Names, job title and contact details of people interviewed:

Pre-interview requests

Please ask the firm to have ready the following information for when you go to interview them:

1. Zesco bills by month / records of Zesco kWh consumed for each month of 2016-present as well as March and August 2014 and 2015
2. Logs of genset usage in terms of diesel fuel used (litres) and hours used for each month 2016-present as well as Feb and July 2014 and 2015
3. Details of generators (kVA, year purchased in)

Questions for accounts

Questions for manufacturing operations

Questions for manufacturing operations that accounts might have

Section A: General information questions for management / accounting

A1. In what year did your firm start manufacturing? ……………………………………

Please note that this does not preclude name changes so long as the company was manufacturing before, as opposed to just trading.

A2. Please fill in the below

Type of employee	Number of employees
Full-time	
Part-time	
Casual	

A3a. Does your firm export? Yes ☐ No ☐

If yes, A3b. please fill below:

Transcontinental?	List of countries
Inside Africa	
Outside Africa	

A4. What were your firm's revenues in the last fiscal year? ZMK ___________

A5. Is your firm majority owned by non-Zambians? Yes ☐ No ☐

A5a. If so, what is the majority nationality owner?

<u>Section B</u>: Energy Consumption, Capacity Utilization and Electricity Supply Performance for engineers/accounting/management

B1bbb. How much did power outages account for loss in production and revenues in the following years?

Year	No loss in production due to power outages (0)	Lost 1-15% of targeted production (1)	Lost 16-30% of targeted production (2)	Lost 31-50% of targeted production (3)	Lost more than 50% of targeted production (4)
2018 Jan – present					
2017					
2016					
2015					
2014					

B1dd. Can you name specific months when power outages have been particularly disruptive to your firm?

B1c. Please share with us records of your firm's on-grid energy consumption (kWh) by month for all months from 2016-present and for March and August 2015 and 2016. **[We would like to take photocopies of your records and not waste your time filling this out in front of you. We would like monthly on-grid energy consumption, in kWh]**

B1cii. Can we request for your on-grid energy consumption (kWh) from Zesco?

B2. What coping strategies does your company adopt for outages when they occur?

	Do not use (0)	**Use a bit (1)**	**Use to a moderate extent (2)**	**Use to a major extent (3)**	**Rely upon all the time (4)**
Voltage regulators and/or use capacitors					
Use surge protectors					
Use power factor correction units					
Reduce output					
Delay production					
Reschedule worker hours (and if so, what percentage of the workforce?)					
Lay-off workers					
Self-generation of power as back-up (and if so, what percentage of normal energy consumption do they make up?) and/or use UPS and/or back-up batteries					
Back-up data system					
Carry larger inventory than otherwise would					
Switch to less energy intensive operation					
Insurance					
Injury to workers					
Other:					

B3a. Please fill out the below table if the company has generators

	Generator 1	Generator 2	Generator 3	Generator 4
i. Financing mechanism 1 Purchased outright 2 Lease 3 Borrow 4 Share 5 Do not possess 6 Other 7 Bought by parent company **If 5, skip to B4a.** Other________				
ii. Capacity, kVA iia. Capacity, litres				
For purchased generators				
iii. Year of purchase				
Variable costs				
v. Fuel, litres/month for all months 2016-present and Feb and July 2014, 2015 vi. Hours of generator use/month for 2 for all months 2016-present and Feb and July 2014, 2015 **[We would like to take photocopies of your records and not waste your time filling this out in front of you]**				

B3b. When the grid power goes out, does your company use all of its available self-generation capacity? Please circle **ONE** of the 4 options below:

0. No, we barely use our generators at all for either planned or unplanned outages.
1. We use our generators to maximum capacity for planned outages, but this does not generate as much energy as we would have used had the grid not gone out. We do not use our generators for unplanned outages.
2. We use our generators to maximum capacity for both planned and unplanned outages, but this does not generate as much energy as we would have used had the grid gone out.
3. Yes, we use our generator(s) at full load, and this helps us replace all the energy we would have used if the grid power had not gone out.

B3bii. If the answer is 1 or 2, does self-generation exclude production? Yes ☐ No ☐

B3e. Would your firm sell electricity back to the grid if there was a mechanism to do so?

Yes ☐ No ☐

B4a. Using your diary, please provide us a picture of the outages and generated hours in the last week **that there were outages -**

Day of the week	**i Production hours** (example: 6am-6pm)	**ii Frequency of planned outages** (example: 2x)	**iii Cumulative planned outages, hours** (example: 2h)	**iv Frequency of unplanned outages** (example: 4x)	**v Cumulative unplanned outages, hours** (example: 0.5h)	**vi Cumulative hours of self-generated power, hours** (example: 2h)
Monday						
Tuesday						
Wednesday						
Thursday						
Friday						
Saturday						
Sunday						

B5a. To what degree do **unplanned** power losses cause each of the following?
B5b. To what degree do **planned** power losses cause each of the following?

Please note that this question is seeking whether there is a difference between the damage between planned and unplanned. Please see next table.

B7. Would you pay a higher tariff if it guaranteed more reliable supply? Yes ☐ No ☐

B8. If yes, up to how many ZMK/kWh on top of the current tariff would it make sense for this firm to pay? ....

Thank you for your time. We will share our overall results with you.

Table for 5a and b.

	Do not cause (0)	**Cause to a minor degree (1)**	**Cause to a moderate degree (2)**	**Cause to a major degree (3)**	**Cause every time (4)**
i.Damage to equipment Unplanned: Planned:					
ii.Damage to inventory Unplanned: Planned: *Please explain how*					
iii.Extra pay to staff to make up for lost productive hours after work hours Unplanned: Planned:					
iv.Damage to firm reputation Unplanned: Planned:					
v.Loss in clients Unplanned: Planned:					
vi.Additional security costs Unplanned: Planned:					
vii. Other 1 Unplanned: Planned:					
viii. Other 2 Unplanned: Planned:					
ix. Other 3 Unplanned: Planned:					

Appendix C: Key Informants with Whom Interviews Were Secured

Government stakeholder interviews

1. ZESCO economists on 10 November 2017 on 6 June 2018 in Lusaka and again on 3 September 2019; and a director on 4 September 2019
2. Department of Energy officers on 6 June 2018 in Lusaka and again on 4 September 2019 as well as correspondence on 30 April 2020
3. Energy Regulation Board Director of Economic Regulation, also Principal Investigator, Impact of Load Shedding on Small Scale Enterprises on 6 June 2018 in Lusaka; his colleague and co-author on 4 September 2019

Private energy developer interviews

4. CEO of a developer of an independent coal power plant on 4 June 2018 in Lusaka
5. CCO Cathy Oxby and CEO Ana Hajduka of Africa GreenCo, a prospective credit-worthy off-take alternative to ZESCO for renewable energy in Zambia on 23 April 2019 and 1 August 2019 September 2019 in London
6. CEO of a developer in Lusaka on 3 September 2019

I. Ahmed, *The Political Economy of Hydropower Dependant Nations*,
https://doi.org/10.1007/978-3-030-71266-2

7. Mikko Marttala, CFO at KPA Unicon, a developer and engineering, procurement, construction contractor of energy solutions using industrial waste on 2 November 2018 in London

Public and private advocates for the manufacturing and private sector

8. Zambia Association of Manufacturers CEO Chipego Zulu on 28 May 2018 in Lusaka
9. Zambia Chamber of Commerce and Industry (ZACCI), Chairman 2004–2007 Wamulume Kalabo
10. Zambia Development Agency M&E Officer on 30 May 2018 in Lusaka

Donors

11. A consultant with 40 years in Zambian and African energy policy working for a donor agency on 27 February 2019 by telephone

Colleague researcher interviews

12. Yimbilanji Sichone, Principal Author of Electricity load shedding: An econometric analysis of the productivity of firms in the manufacturing sector in Lusaka on 26 May 2018 in Lusaka
13. IGC Zambia Country Economist Miljan Sladoje on 28 May 2018 in Lusaka

Other useful unstructured conversations

14. Chibelushi Maxwell Musongole, Lecturer and retired Assistant Director at the Central Bank of Zambia on 28 May 2018 in Lusaka
15. Natty Chilundiki, Research Coordinator, on 29 May 2018 in Lusaka
16. Jim Friedlander, a lawyer with decades of experience working in Africa, on 3 May 2018 in London
17. Michael Mainelli, who as a consulting partner with BDO Binder Hamlyn, looked at the energy supply to Zambia Consolidated Copper Mines in 1992–1993. Email on 18 February 2019

We also visited the following institutions for requests for data:

18. Patents and Companies Registration Agency (PACRA)
19. Bank of Zambia on 30 May 2018
20. Ministry of Commerce, Trade and Industry on 30 May and 6 June 2018
21. Central Statistical Office on 30 May and 6 June 2018
22. Zambia Meteorological Department on 26 October 2018

References

Adenikinju, A. F. (2003). Electric infrastructure failures in Nigeria: A survey-based analysis of the costs and adjustment responses. *Energy Policy, 31*(14), 1519–1530. https://doi.org/10.1016/S0301-4215(02)00208-2

AfDB. (2013). *AfDB approves US $150-million loan to support access to reliable electricity supply in Zambia, AfDB website*. Retrieved May 25, 2020, from https://www.afdb.org/en/news-and-events/afdb-approves-us-150-million-loan-to-support-access-to-reliable-electricity-supply-in-zambia-12334

Africa Power Ventures. (2018). *SAPP market analysis*. Ebene.

African Energy. (2014, May 30). Zambia: Maamba plant delayed by transmission hold-up. *African Energy Newsletter*. https://www.africa-energy.com/article/zambia-maamba-plant-delayed-transmission-hold

African Energy. (2019, March). Zambia: Zesco seeks big tariff increase. *African Energy*.

Ahmed, I. (2017, March 27). Leapfrogging into the light. *Financial Times*. https://www.ft.com/content/41524878-108f-11e7-b030-768954394623

Ahmed, I. (2019). Keynote presentation. In IGC (Ed.), *Nurturing the manufacturing sector: Ensuring strong roots for Zambian growth*. International Growth Centre.

Ahmed, I. (2020a). Explaining Rwanda's prioritisation of rural electrification over rural clean drinking water through institutional path dependency. *Structural Change and Economic Dynamics, 54*(September), 186–201. https://doi.org/10.1016/j.strueco.2020.05.001

I. Ahmed, *The Political Economy of Hydropower Dependant Nations*, https://doi.org/10.1007/978-3-030-71266-2

Ahmed, I. (2020b). The cost of power outages to Zambia's manufacturing firms, households and the climate. In *Grid reliability and utility operations*. The Applied Research Programme on Energy and Economic Growth (EEG).

Ahmed, I. (2021). *The political economy of energy mix in hydropower dependent developing nations–a case study of Zambia*. University College London. Available at: https://discovery.ucl.ac.uk/id/eprint/10121236/.

Ahmed, I., Baddeley, M., Coffman, D., Meikle, J., Oseni, M., & Sianjase, G. (2019). *The cost of power outages to Zambia's manufacturing sector*. https://www.theigc.org/wp-content/uploads/2019/08/Ahmed-et-al-2019-policy-brief-41408.pdf. https://doi.org/F-41408-ZMB-1

Ahmed, I., Parikh, P., Sianjase, G., & Coffman, D. (2020, December 4). The impact decades long dependence on hydropower in El Niño impact-prone Zambia is having on carbon emissions through diesel backup generation. *Environmental Research Letters, 15*(12) https://doi.org/10.1088/1748-9326/abb6a1

Al Jazeera. (2011, September 23). Opposition leader wins Zambia election. *Al Jazeera*. https://www.aljazeera.com/news/2011/9/23/opposition-leader-wins-zambia-election

Alacevich, M. (2009). *The political economy of the World Bank—The early years*. Stanford University Press.

Allcott, H., Collard-Wexler, A., & O'Connell, S. D. (2015). *How do electricity shortages affect industry? Evidence from India, NBER Working Paper Series* 19977, Cambridge, MA. https://doi.org/10.1257/aer.20140389

Andersen, T. B., & Dalgaard, C. J. (2013). Power outages and economic growth in Africa. *Energy Economics, 38*, 19–23. https://doi.org/10.1016/j.eneco.2013.02.016

Arnold, J. M., Mattoo, A., & Narciso, G. (2008). Services inputs and firm productivity in Sub-Saharan Africa: Evidence from firm-level data. *Journal of African Economies, 17*(4), 578–599. https://doi.org/10.1093/jae/ejm042

Arup. (2017). *ASPIRE—A Sustainability Poverty and Infrastructure Routine for Evaluation*. https://sustainabilitytoolkit.arup.com/#/.

Baddeley, M. (2003). *Investment: Theories and analysis*. Palgrave Macmillan.

Baddeley, M. (2013). Behavioural economics and finance. First edit. Routledge.

Bakker, K. (2013). Constructing 'public' water: the World Bank, urban water supply, and the biopolitics of development. *Environment and Planning D: Society and Space, 31*, 280–300. https://doi.org/10.1068/d5111

Bank of Zambia, Central Statistical Office and Zambia Development Agency. (2017). *Foreign Private Investment & Investor Perceptions in Zambia—2017: Enhancing Investment for Export Promotion towards Inclusive Growth*. Government of the Republic of Zambia.

Barber, W. J. (1996). *Designs within disorder. Franklin D. Roosevelt, the Economists, and the Shaping of American Economic Policy, 1933–1945*. Cambridge University Press.

Barrick Gold Corporation. (2019). *Lumwana, Barrick website.* Retrieved October 2, 2019, from https://www.barrick.com/operations/lumwana/default.aspx

Barton, S. J. (2014). Sino-substitution: Chinese foreign direct investment in Zambia. *Journal of Chinese Economic and Foreign Trade Studies, 7*(2), 90–109. https://doi.org/10.1108/JCEFTS-08-2013-0025

Barton, S. J. (2016). *Policy signals and market responses—A 50 year history of Zambia's relationship with foreign capital.* (A. B. Bell, D. Coffman, & T. K. Moore, Eds.), 1st ed. Palgrave Macmillan.

Batidzirai, B., Moyo, A., & Kapembwa, M. (2018). *Willingness to pay for improved electricity supply reliability in Zambia—A survey of urban enterprises in Lusaka and Kitwe.* IGC.

Bayliss, K., & Pollen, G. (2019). *From Global to Local: Exploring the effects of the global energy policy paradigm in Zambia.* 119. Leeds. https://sri-working-papers.leeds.ac.uk/wp-content/uploads/sites/67/2019/10/SRIPs-119.pdf

BBC. (2008). *Obituary: Zambia's Levy Mwanawasa, British Broadcasting Corporation News.* Retrieved November 23, 2018, from http://news.bbc.co.uk/1/hi/world/africa/7488623.stm

Beck, L., & Bernauer, T. (2011). How will combined changes in water demand and climate affect water availability in the Zambezi river basin? *Global Environmental Change, 21*(3), 1061–1072. https://doi.org/10.1016/j.gloenvcha.2011.04.001

Becker, M. C., et al. (2005). Applying organizational routines in understanding organizational change. *Industrial & Corporate Change, 14*(5), 775–791. https://doi.org/10.1093/icc/dth071

Beenstock, M., Goldin, E., & Haitovsky, Y. (1997). The cost of power outages in the business and public sectors in Israel: Revealed preference vs. Subjective valuation. *The Energy Journal, 18*(2), 39–61. https://www.jstor.org/stable/41322730

Beilfuss, R. (2012). *A risky climate for Southern African Hydro: Assessing hydrological risks and consequences for Zambezi River Basin Dams.* International Rivers. http://seine.oieau.fr/oieau.fr/sites/seine.oieau.fr.oieau.fr.eaudoc/files/32987.pdf

BEIS (2019). 2019 Government Greenhouse Gas Conversion Factors for Company Reporting - Methodologhy Paper for Emission Factors. London. Available at: https://assets.publishing.service.gov.uk/government/uploads/system/uploads/attachment_data/file/829336/2019_Green-house-gas-reporting-methodology.pdf.

Bental, B., & Ravid, A. S. (1982). A simple method for evaluating the marginal cost of unsupplied electricity. *The Bell Journal of Economics, 13*(1), 249–253. http://www.jstor.org/stable/3003447

Bhorat, H. et al. (2017). *Understanding growth-income inequality interactions in Zambia.* https://www.theigc.org/wp-content/uploads/2018/04/Bhorat-et-al-2017-final-paper.pdf

Boos, A., & Holm-Müller, K. (2016). The Zambian Resource Curse and its influence on Genuine Savings as an indicator for 'weak' sustainable development. *Environment, Development and Sustainability, 18*(3), 881–919. https://doi.org/10.1007/s10668-015-9667-5

Boyden, A., Kie, V., & Doolan, M. (2016). The environmental impacts of recycling portable lithium-ion batteries. *Procedia CIRP, 48*, 188–193. https://doi.org/10.1016/j.procir.2016.03.100

BRE Global. (2019). *CEEQUAL Version 6 technical manual.* BRE Global. https://www.ceequal.com/version-6/

British American Tobacco. (2016). *Audited financial results for the year ended 31 December 2015.*

British American Tobacco. (2017). Annual report & financial statements for the year ended 31 December 2016. Lusaka.

British American Tobacco. (2018). Annual report & financial statements for the year ended 31 December 2017. Lusaka.

Buragohain, T. (2012). Impact of solar energy in rural development in India. *International Journal of Environmental Science and Development, 3*(4), 334–338. http://www.ijesd.org/papers/242-B10021.pdf

Cai, W., McPhaden, M. J., Grimm, A. M., et al. (2020). Climate impacts of the El Niño–Southern Oscillation on South America. *Nature Reviews Earth & Environment, 1*, 215–231. https://doi.org/10.1038/s43017-020-0040-3

Cantore, N., et al. (2017). Manufacturing as an engine of growth: Which is the best fuel? *Structural Change and Economic Dynamics, 42*, 56–66. https://doi.org/10.1016/j.strueco.2017.04.004

Cardozo, A., et al. (2014). Growth, employment, diversification, and the political economy of private sector development in Zambia. In C. Adam, P. Collier, & M. Gondwe (Eds.), *Zambia: Building prosperity from resource wealth.* Oxford Scholarship Online. https://doi.org/10.1093/acprof:oso/9780199660605.003.0003

Castellano, A. et al. (2015). *Brighter Africa: The growth potential of the sub-Saharan electricity sector.* https://www.mckinsey.com/~/media/McKinsey/dotcom/client_service/EPNG/PDFs/Brighter_Africa-The_growth_potential_of_the_sub-Saharan_electricity_sector.ashx.

Cecere, G., et al. (2014). Lock-in and path dependence: an evolutionary approach to eco-innovations. *Journal of Evolutionary Economics, 24*(5), 1037–1065. https://doi.org/10.1007/s00191-014-0381-5

Chen, C.-Y., & Vella, A. (1994). Estimating the economic costs of electricity shortages using input-output analysis: the case of Taiwan. *Applied Economics, 26*(11), 1061–1069. https://doi.org/10.1080/00036849400000122

Christensen, C. (1997). *The innovator's dilemma.* Harvard Business School Press.

CIA. (2016). *Zimbabwe, The World Factbook.*

Circular Energy Storage. (2019). *The lithium-ion battery end-of-life market-a baseline study, global battery alliance.* http://www3.weforum.org/docs/GBA_EOL_baseline_Circular_Energy_Storage.pdf

Coffman, D. (2013a). Credibility, transparency, accountability, and the public credit under the Long Parliament and Commonwealth, 1643–1653. In D. Coffman, A. Leonard, & L. Neal (Eds.), *Questioning Credible Commitment—Perspectives on the rise of financial capitalism.* Cambridge University Press.

Coffman, D. (2013b). Towards a New Jerusalem: The committee for regulating the excise, 1649–1653 Author (s): D'Maris Coffman Source: The, December 2013, Vol. 128, No. 535 (December Stable Url: http://www.jstor.com/stable/24473893 Towa. English Historical Review, *128*(535), 1418–1450. https://doi.org/10.1093/ehr/cet274

Coffman, D., Leonard, A., & Neal, L. (2013). *Questioning credible commitment—Perspectives on the rise of financial capitalism* (D. Coffman, A. Leonard, & L. Neal, Eds.). Cambridge University Press.

Coffman, D., & Neal, L. (2013). Introduction. In D. Coffman, A. Leonard, & L. Neal (Eds.), *Questioning credible commitment—Perspectives on the rise of financial capitalism.* Cambridge University Press.

Coffman, D. et al. (2020). Ecological transition. In Cerniglia, F. and Saraceno, F. (eds.), *A European public investment outlook*, 161–174. https://doi.org/10.11647/OBP.0222

Colignon, R. A. (1997). *Power plays: Critical events in the institutionalization of the Tennessee valley authority.* State University of New York Press.

Conway, D., et al. (2015, September). Climate and southern Africa's water–energy–food nexus. *Nature Climate Change*. Nature Publishing Group, 5, 837–846. https://doi.org/10.1038/nclimate2735

Conway, D., et al. (2017). Hydropower plans in eastern and southern Africa increase risk of concurrent climate-related electricity supply disruption. *Nature Energy, 2*(12), 946–953. https://doi.org/10.1038/s41560-017-0037-4

Cooke, M. L. (1949, July 11). Down to Earth with Point Four. *New Republic*, 18–21.

Cornish, J.-J. (2019, August 24). South Africa scraps Russian nuclear plant plans. *Radio France International.* http://www.rfi.fr/en/africa/20190824-south-africa-cancel-russia-nuclear-plant-order-economy-ramaphosa

Costa Rica Guide. (2020). *La Nada, La Niña & El Niño in Costa Rica*, *Costa Rica Guide.* Retrieved October 17, 2020, from https://costa-rica-guide.com/travel/weather/la-nada-la-nina-el-nino-in-costa-rica/#:~:text=El Niño was partially responsible,drought from 2015 through 2016.&text=ENSO has the biggest impacts,season in June through December.

CSO. (2018). *Data portal tables.* Central Statistical Office of Zambia.

CSO and Ministry for Labour and Social Security. (2019). 2018 Labour Force Survey Report. *Ministry of Labour and Social Security*, p. 153. http://www.

statistics.gov.my/portal/download_Labour/files/labour_force/Labour_Force_Survey_Report_Malaysia_2012.pdf

Currie, L. (1950). *The basis of a development program for Colombia*. https://doi.org/10.2105/ajph.40.12.1567-a

Currie, L. B. (1979, August 31). *A talk by Lauchlin Currie: Hotel Tequendama*. Mimeo.

Dalberg Advisors, Lighting Global and World Bank. (2018). *Off-Grid Solar Market Trends Report 2018*. Washington, DC. https://doi.org/10.1017/CBO9781107415324.004

Davies, A., & Brady, T. (2000). Organisational capabilities and learning in complex product systems: Towards repeatable solutions. *Research Policy, 29*(7–8), 931–953. https://doi.org/10.1016/s0048-7333(00)00113-x

Deemer, B. R., et al. (2016). Greenhouse gas emissions from reservoir water surfaces: A new global synthesis. *Bioscience, 66*(11), 949–964. https://doi.org/10.1093/biosci/biw117

Díaz, J., Lüders, R., & Wagner, G. (2016). *Chile 1810–2010: La República En Cifras: Historical statistics*. First. Ediciones Universidad Católica of Chile. https://www.jstor.org/stable/j.ctt1h9cvqj

Diboma, B. S., & Tamo Tatietse, T. (2013). Power interruption costs to industries in Cameroon. *Energy Policy, 62*, 582–592. https://doi.org/10.1016/j.enpol.2013.07.014

Diggers Editor. (2019, November 20). A history of debt, poor management of Zesco. *Diggers*. https://diggers.news/opinion/2019/11/20/a-history-of-debt-poor-management-at-zesco/

Dlamini, J., et al. (2016). Relationship between energy consumption and economic growth in South Africa: Evidence from the bootstrap rolling-window approach. *Energy Sources, Part B: Economics, Planning and Policy, 11*(7), 617–625. https://doi.org/10.1080/15567249.2013.843618

DNV GL. (2020). *Energy transition outlook 2020*. DNV GL.

Döll, P., et al. (2018). Risks for the global freshwater system at 1.5°C and 2°C global warming. *Environmental Research Letters, 13*. https://doi.org/10.1088/1748-9326/aab792

Dosi, G. (1982). Technological paradigms and technological trajectories. A suggested interpretation of the determinants and directions of technical change. *Research Policy, 11*(3), 147–162. https://doi.org/10.1016/0048-7333(82)90016-6

Dosi, G., Teece, D., & Al, E. (1992). Toward a theory of corporate coherence: Some preliminary remarks. In G. Dosi, R. Giannetti, & P. Toninelli (Eds.), *Technology and enterprise in a historical perspective*. Clarendon Press.

Douglas, W. O. (1951). Address before the General Assembly of the State of Tennessee. Appendix to the Congressional Record, 82nd Congress, 1 sess., p. 22 February.

Eberhard, A., et al. (2008, May). Underpowered: The State of the Power Sector in Sub-Saharan Africa. *Background Paper 6, Africa Infrastructure Country Diagnostic*, p. 54. https://doi.org/10.1109/TENCON.2008.4766817

Ekbladh, D. (2002). 'Mr. TVA': Grass-roots development, David Lilienthal, and the rise and fall of the Tennessee Valley authority as a symbol for U.S. overseas development, 1933–1973. *Diplomatic History, 26*(3), 335–374. https://doi.org/10.1111/1467-7709.00315

Energy Regulation Board. (2011). *Energy Sector Report 2010.*

Energy Regulation Board of Zambia. (2015). *Energy Regulation 2014.*

Energy Regulation Board of Zambia. (2016a). *Energy Sector Report 2015.* http://www.erb.org.zm/reports/EnergySectorReport2014.pdf

Energy Regulation Board of Zambia. (2016b). Request for expressions of interest: Electricity cost of service study and determination of economic cost-based tariffs. African Development Bank Group. https://www.afdb.org/fileadmin/uploads/afdb/Documents/Procurement/Project-related-Procurement/EOI_-_Zambia_-_Consulting_services_-_Electricity_Cost_of_Service_Study_and_Determination_of_Economic_Cost-based_Tariffs.pdf

Energy Regulation Board of Zambia. (2017a). *Energy Sector Report 2016.* https://doi.org/10.13140/RG.2.1.3893.7043

Energy Regulation Board of Zambia. (2017b). *Public Notice, April 2017.* http://www.erb.org.zm/downloads/zescoPCP2017FinalVersionDraft.pdf

Energy Regulation Board of Zambia. (2018). *Energy Sector Report 2017.* http://www.erb.org.zm/downloads/reports/esr2017.pdf

Energy Regulation Board of Zambia. (2019a). *Energy Sector Report 2018.* http://www.erb.org.zm/reports/esr2018.pdf

Energy Regulation Board of Zambia. (2019). Public notice—Board's decision on ZESCO's application to adjust electricity tariffs and connection charges for the year 2019.

Escribano, A., Guasch, J. L., & Pena, J. (2010, January). Assessing the impact of infrastructure quality on firm productivity in Africa: Cross-country comparisons based on investment climate surveys from 1999 to 2005, p. 117. https://doi.org/10.1596/1813-9450-5191

ESI-Africa. (2019, November). Zimbabwe includes pumped storage hydro in renewable energy strategy. *Hydro Review*. https://www.hydroreview.com/2019/11/18/zimbabwe-includes-pumped-storage-hydro-in-renewable-energy-strategy/#gref

European Commission. (2015). *Commission decision financing humanitarian actions in Central Africa from the 11th European Development Fund (EDF).* https://ec.europa.eu/echo/sites/echo-site/files/1_EN_autre_document_appui_part1_v1_0.pdf

Falchetta, G., et al. (2019). Hydropower dependency and climate change in sub-Saharan Africa : A nexus framework and evidence-based review. *Journal of Cleaner Production, 231*, 1399–1417. https://doi.org/10.1016/j.jclepro.2019.05.263

FAO. (2015a). *El Niño in Asia: Prolonged dry weather in several countries affecting plantings and yield potential of the 2015 main season food crops.* http://www.fao.org/3/a-I4829E.pdf

FAO. (2015b). *FAO and El Niño, Food and Agriculture Oganization of the United Nations.* Retrieved March 28, 2020, from http://www.fao.org/el-nino/en/

FAO. (2016). *2015–2016 El Niño: Early action and response for agriculture, food security and nutrition.* http://www.fao.org/3/a-i6049e.pdf.

FAO. (2018). *El Niño, Food and Agriculture Oganization of the United Nations.* Retrieved March 28, 2020, from http://www.fao.org/emergencies/crisis/elnino-lanina/intro/en/

FAO. (2020). *Aquastat database, Aquastat database.* Retrieved December 28, 2020, from http://www.fao.org/nr/water/aquastat/data/query/results.html

Farquharson, D., Jaramillo, P., and Samaras, C. (2018, October). Sustainability implications of electricity outages in sub-Saharan Africa', Nature Sustainability, 1, 589–597. https://doi.org/10.1038/s41893-018-0151-8 Sustainability.

Finer, H. (1972). *The T.V.A.: Lessons for international application.* Da Capo Press.

First Quantum Minerals Ltd. (2019). *Kansanshi, first quantum minerals website.* Retrieved October 2, 2019, from https://www.first-quantum.com/Our-Business/operating-mines/Kansanshi/

Fisher-Vanden, K., Mansur, E. T., & Wang, Q. J. (2015). Electricity shortages and firm productivity: Evidence from China's industrial firms. *Journal of Development Economics, 114*, 172–188. https://doi.org/10.1016/j.jdeveco.2015.01.002

Ford, N. (2016, August). Gas fuels Tanzania, Mozambique growth. *The Africa Report.* https://www.theafricareport.com/939/gas-fuels-tanzania-mozambique-growth/

Foster, V., & Steinbuks, J. (2009). *Paying the price for unreliable power supplies—In-house generation of electricity by firms in Africa, The World Bank Policy Research Working Paper.* https://doi.org/10.1596/1813-9450-4913

Fountain, H. (2019, May 17). What Panama's worst drought means for its canal's future. *The New York Times.* https://www.nytimes.com/2019/05/17/climate/drought-water-shortage-panama-canal.html#:~:text=A severe drought in Panama,can safely navigate the waterway.&text=The drought is linked to,to continue into the fall.

Fukuyama, F. (2018). *Identity: The demand for dignity and the politics of resentment.* Macmillan.

Funder, M., et al. (2018). The politics of climate change adaptation in development: Authority, resource control and state intervention in rural Zambia. *The Journal of Development Studies. Routledge, 54*(1), 30–46. https://doi.org/10.1080/00220388.2016.1277021

Garner, R. L. (1952). *Letter to Rafael Delgado*. LBCP.

Garner, R. L. (1953). *Letter to Emilio Toro*. LBCP.

GHG Protocol. (2013). *Scope 3 calculation guidance, GHG protocol website*. Retrieved June 28, 2020, from https://ghgprotocol.org/scope-3-technical-calculation-guidance

Gibbs, J. (2019). *Planet of the humans*. Rumble Media and YouTube. https://www.youtube.com/watch?v=HQrZUj4-Xtk

Gilpin, R. (2001). *Global political economy: Understanding the international economic order*. Princeton University Press.

Global Battery Alliance. (2019). *A vision for a sustainable battery value chain in 2030: Unlocking the full potential to power sustainable development and climate change mitigation*. http://www3.weforum.org/docs/WEF_A_Vision_for_a_Sustainable_Battery_Value_Chain_in_2030_Report.pdf

Gómez, D. R., et al. (2006). Chapter 2 stationary combustion. In A. Garg & T. Pulles (Eds.), *2006 IPCC guidelines for national greenhouse gas inventories Vol. 2 energy*. Institute for Global Environmental Strategies.

Gondwe, M., & Pamu, E. M. (2014). Growth strategies for Zambia: Harnessing natural resource wealth for sustainable growth. In C. Adam, P. Collier, & M. Gondwe (Eds.), *Zambia: Building prosperity from resource wealth*. Oxford Scholarship Online. https://doi.org/10.1093/acprof:oso/9780199660605.003.0001

Government of Zambia. (2008). *The customs and excise (electrical machinery and equipment) (suspension) regulations*.

Government of Zambia. (2016). Zambia's intended nationally determined contribution to the 2015 agreement on climate change. UNFCCC NDC Registry. https://www4.unfccc.int/sites/ndcstaging/PublishedDocuments/ZambiaFirst/FINAL+ZAMBIA'S+INDC_1.pdf

Gunaratne, L., & LGA Consultants. (2002). *Overview of the electricity sector in relation to public private partnerships in Sri Lanka*. https://assets.publishing.service.gov.uk/media/57a08d4040f0b6497400174e/R8148-Ethioprep.pdf

Hall, P. A. (2010). Historical institutionalism in rationalist and sociological perspective. In J. Mahoney & K. Thelen (Eds.), *Explaining institutional change—Ambiguity, agency and power*. Cambridge University Press.

Hannan, M., & Freeman, J. (1989). *Organizational ecology*. Harvard University Press.

Hao, Z., et al. (2018, September). Quantifying the relationship between compound dry and hot events and El Niño—Southern Oscillation (ENSO) at the global scale. *Journal of Hydrology, 567*, 332–338. https://doi.org/10.1016/j.jhydrol.2018.10.022

Haraguchi, N., Martorano, B., & Sanfilippo, M. (2019). What factors drive successful industrialization? Evidence and implications for developing countries. *Structural Change and Economic Dynamics, 49*, 266–276. https://doi.org/10.1016/j.strueco.2018.11.002

Haria, S., & Ahmed, I. (2020). *No Increasing tariffs to prevent another electricity crisis in Zambia, International Growth Centre blog*. Retrieved May 7, 2020, from https://www.theigc.org/blog/increasing-tariffs-to-prevent-another-electricity-crisis-in-zambia/

Harmon, K. (2009, June 29). How does geothermal drilling trigger earthquakes? *Scientific American*. https://www.scientificamerican.com/article/geothermal-drilling-earthquakes/

Harrison, G. P., & Whittington, H. W. (2002). Susceptibility of the Batoka Gorge hydroelectric scheme to climate change. *Journal of Hydrology, 264*(1–4), 230–241. https://doi.org/10.1016/S0022-1694(02)00096-3

Henderson, R. M., & Clark, K. B. (1990). Architectural innovation: The reconfiguration of existing product technologies and the failure of established firms. Administrative Science Quarterly, *35*(Technology, Organizations and Innovation), 9–30. https://doi.org/10.2307/2393549

Hirschman, A. O. (1959). *The strategy of economic development*. Yale University Press.

Huxley, J. (1946). *TVA—Adventure in planning*. The Architectural Press.

IBRD. (1947). *Second Annual Report 1946–1947*.

IBRD. (1948). *Third Annual Report, 1947–1948*. https://doi.org/10.1007/978-1-349-26356-1_7

IBRD. (1950a). *Report and recommendations of the President to the Executive Directors on the proposed loan to Central Hidroelectrica de Caldas*. http://documents1.worldbank.org/curated/en/657441468022430739/pdf/multi0page.pdf

IBRD. (1950b). *Report and recommendations of the President to the Executive Directors on the proposed loan to Central Hidroelectrica del Rio Anchicaya*. Retrieved from http://documents1.worldbank.org/curated/en/619771468240592149/pdf/multi0page.pdf

IBRD. (1950c). *Technical report of the Paulo Afonso Hydroelectric Project in Brazil*. https://projects.worldbank.org/en/projects-operations/projects-list?lang=en&searchTerm=&countrycode_exact=BR

IBRD. (1951a). *Report and recommendations of the President to the Executive Directors on the proposed loan to Central Hidroelectrica del Rio Lebrija*. http://documents1.worldbank.org/curated/en/216361468026996331/pdf/multi0page.pdf

IBRD. (1951b). *Report on Cuba—Findings and recommendations of an economic and technical mission*. https://doi.org/10.2307/2509624

IBRD. (1951c). *Sixth Annual Report 1950–1951*.

IBRD. (1952a). *Technical report on the Southern Rhodesian power expansion project*.

IBRD. (1952b). *The Seyhan River multipurpose project in Turkey*. http://documents1.worldbank.org/curated/en/855141468120269687/pdf/multi-page.pdf

IBRD. (1953a). *Report and recommendations of the President to the Executive Directors on the proposed loan to the Companhia de Electricidade do Alto Rio Grande (CEARG) and to the Centrais Eletricas de Minas Gerais S.A. (cemig) in the United States of Brazil (Itutinga Hy*. https://projects.worldbank.org/en/projects-operations/projects-list?lang=en&searchTerm=&countrycode_exact=BR

IBRD. (1953b). *The economic development of Nicaragua*.

IBRD. (1956a). *$5 million loan in Ecuador*. http://documents1.worldbank.org/curated/en/881521589352634484/pdf/Announcement-of-Five-Million-Dollars-Loan-in-Ecuador-on-March-29-1956.pdf

IBRD. (1956b). *Appraisal of Baygorria hydroelectric project Uruguay*. http://documents1.worldbank.org/curated/en/461381468317969305/pdf/multi-page.pdf

IBRD. (1956c). *Appraisal of the Kariba Hydroelectric Project in the Federation of Rhodesia and Nyasaland*. https://projects.worldbank.org/en/projects-operations/projects-list?countrycode_exact=ZW

IBRD. (1959). *Appraisal of an interim power project for Honduras*. http://documents1.worldbank.org/curated/en/160141468254056902/pdf/multi-page.pdf

IBRD. (1960a). *Appraisal of Huinco Hydroelectric Project Lima Light and Power Company (Empresas Elect Ricas Asociadas) Peru*. http://documents1.worldbank.org/curated/en/754051468145164676/pdf/multi0page.pdf

IBRD. (1960b). *Appraisal of The Canaveral Hydroelectric Project (First Stage of the Lake Yojoa—Rio Lindo Hydroelectric Development)*. http://documents1.worldbank.org/curated/en/887431468251089233/pdf/multi0page.pdf

IBRD. (1961a). *Appraisal of The Volta River Hydroelectric Project Ghana*. http://documents1.worldbank.org/curated/en/684761468030834258/pdf/multi-page.pdf

IBRD. (1961b). *Report and recommendations of the President to the Executive Directors on a proposed loan public disclosure authorized to the Instituto Costarricense de Electricidad in the Republic of Costa Rica*. http://documents1.worldbank.org/curated/en/357131468234862067/pdf/multi0page.pdf

IBRD. (1961c). *Report and recommendations of the President to the Executive Directors on a proposed loan to the Ethiopian electric light and power authority*. https://doi.org/10.1017/S0020818300024826

IBRD. (1962). *Appraisal of the electrification of the Central Provinces Panama*. http://documents1.worldbank.org/curated/en/396891468286467077/pdf/multi-page.pdf

IBRD. (1967). *The Instituto Nacional de Electrificacion—The Jurun-Marinala hydroelectric scheme, associated transmission lines and other generating facilities, Guatemala.* http://documents1.worldbank.org/curated/en/720171468251383318/pdf/multi0page.pdf

IBRD. (2020). *Subscriptions and voting power of member countries.* http://pubdocs.worldbank.org/en/795101541106471736/IBRDCountryVotingTable.pdf

IBRD and IDA. (1961). *Appraisal of the Sudan Power Project Central Electricity and Water Corporation (Cewc) Sudan, Appraisal Report.* https://doi.org/10.1017/S0020818300024826

IBRD and IDA. (1964). *Central African Power Corporation Project Rhodesia.* https://projects.worldbank.org/en/projects-operations/projects-list?countrycode_exact=ZW

IBRD and IDA. (1968). *The Instituto Nacional De Electrificacion Guacalate Power Project Guatemala.* https://doi.org/10.1017/S0020818300024826

IBRD and IDA. (1970). *Appraisal of Kariba North Hydroelectric Project Zambia.* http://documents.worldbank.org/curated/en/917071468334477402/pdf/multi-page.pdf

IBRD and IDA. (1971). *Appraisal of the Kamburu Hydroelectric Project Tana River Development Company Limited Republic of Kenya.* https://doi.org/10.1017/S0020818300024826

IBRD and IDA. (1973). *Appraisal of Kafue Hydroelectric Project (Stage II) Zambia.* http://documents.worldbank.org/curated/en/824421468167973026/pdf/multi-page.pdf

IDA. (1964). *Report and recommendations of the President to the Executive Directors on two proposed development credits to the Republic of Bolivia.* http://documents1.worldbank.org/curated/en/190231468208484113/pdf/multi-page.pdf

IEA. (2020). *Africa Energy Outlook 2019—Analysis Scenarios, Africa Energy Outlook 2019*, Paris. https://www.iea.org/reports/africa-energy-outlook-2019#energy-access; https://www.iea.org/reports/africa-energy-outlook-2019#africa-case

IEA. (2020b). *Global EV Outlook 2020, Global EV Outlook.* https://doi.org/10.1787/9789264302365-en

IMF. (2008). *Regional economic outlook—Sub-Saharan Africa.* International Monetary Fund.

IMF. (2019). *Zambia: Gross domestic product, constant prices.* World Economic Outlook. https://www.imf.org/external/pubs/ft/weo/2019/02/weodata/weorept.aspx?sy=2017&ey=2024&scsm=1&ssd=1&sort=country&ds=.&br=1&pr1.x=49&pr1.y=11&c=754&s=NGDP_R&grp=0&a=

International Growth Centre. (2018). *About—IGC, www.theigc.org.* Retrieved November 22, 2018, from https://www.theigc.org/about/

Investment Mine. (2018). *Historical Copper Prices and Price Chart, infomine.com.* Retrieved November 28, 2018, from http://www.infomine.com/investment/metal-prices/copper/all/

IPCC. (2000). *Summary for policymakers—Emissions scenarios.* Intergovernmental Panel on Climate Change. https://www.ipcc.ch/site/assets/uploads/2018/03/sres-en.pdf

IPCC. (2001). *Climate change 2001: Impacts, adaptation, and vulnerability.* https://www.ipcc.ch/site/assets/uploads/2018/03/WGII_TAR_full_report-2.pdf

IPCC. (2018). *Global warming of 1.5°C. An IPCC Special Report on the impacts of global warming of 1.5°C above pre-industrial levels and related global greenhouse gas emission pathways, in the context of strengthening the global response to the threat of climate change, Special Report.* https://www.ipcc.ch/site/assets/uploads/sites/2/2019/06/SR15_Full_Report_Low_Res.pdf

Isango, E. (2017, May 10). DRC Faces Power Shortage Caused by Drought. *Voice of America.* https://www.voanews.com/africa/drc-faces-power-shortage-caused-drought

Japan International Cooperation Agency (JICA) and JERA Co. Inc. (2017). *Data collection survey on Southern African power pool final report (Appendix).*

Jaramillo, P., & Skoknic, E. (1973). *Costo Social de las Restricciones de Energia Electrica.* Oficina de Planificacion, Chile, August.

Jerven, M. (2013). *Poor numbers how we are misled by African development statistics and what to do about it.* Cornell University Press (Cornell studies in political economy).

Jiang, L., et al. (2020). Rural household energy consumption of farmers and herders in the Qinghai-Tibet Plateau. *Energy, 192*, 116649. https://doi.org/10.1016/j.energy.2019.116649

JICA. (2008). *Project formation study for power development planning in Zambia.* https://doi.org/10.1007/s00445-011-0503-y

Johnson, L. B. (1965). *Peace without conquest.* C-SPAN. https://www.c-span.org/video/?153274-1/reel-america-president-lyndon-johnson-vietnam-war-1965

Jourdan, P. P. (1990). *Strategies for the regional planning of the minerals industry in Southern Africa: The case of the SADCC.* University of Leeds. http://etheses.whiterose.ac.uk/490/1/uk_bl_ethos_291037.pdf

Jury, M. (2002). Economic impacts of climate variability in South Africa and development of resource. *American Meteorological Society, 41*(1), 46–55. https://www.jstor.org/stable/26184931

Kahneman, D., & Tversky, A. (1979). Prospect theory: An analysis of decision under risk. *Econometrica, 47*(2), 263–292. https://www.jstor.org/stable/1914185

Kaldor, N. (1970). The case for regional policies. *Scottish Journal of Political Economy, 17*(3), 337–348. https://doi.org/10.1111/j.1467-9485.1970.tb00712.x

Kaldor, N. (1984) *Causes of growth and stagnation in the world economy.* Cambridge University Press (Raffaele Mattioli Series). https://doi.org/10.1017/CBO9780511559709

Kapur, D., Lewis, J. P., & Webb, R. C. (1997). *The World Bank: Its first half century.* Brookings Institution Press. http://documents.worldbank.org/curated/en/313081468322727631/History

Kessides, C. (1993). *The contributions of infrastructure to economic development—A review of experience and policy implications.* The World Bank.

Konkola Copper Mines Plc. (2019). *Corporate Profile, Konkola Copper Mines website.* Retrieved October 2, 2019, from http://kcm.co.zm/corporate-profile/

Kuznets, S. (1955). Economic growth and income inequality. *The American Economic Review, 45*(1), 1–28. https://www.jstor.org/stable/1811581

Lafarge Zambia. (2016). Annual Report 2015. Lusaka.

Lafarge Zambia. (2017). Annual Report 2016. Lusaka.

Lafarge Zambia. (2018). Annual Report 2017. Lusaka.

Lakhani, N. (2019). 'People are dying': how the climate crisis has sparked an exodus to the US. *The Guardian.* https://www.theguardian.com/global-development/2019/jul/29/guatemala-climate-crisis-migration-drought-famine

Lazard. (2018). *Lazard's annual levelized cost of energy analysis—Version 12.0.* www.lazard.com

Levitt, B., & March, J. G. (1988). Organizational learning. *Annual Review of Sociology, 14,* 319–340. http://www.jstor.com/stable/2083321

Liebowitz, S. J., & Margolis, S. E. (1995). Path dependence, lock-in, and history. *Journal of Law, Economics, and Organization, 11*(1), 205–226. https://doi.org/10.2139/ssrn.1706450

Lighting Global, et al. (2020). *Off-grid solar market trends report 2020.* https://doi.org/10.1017/CBO9781107415324.004

Lighting Global and Vivid Economics. (2019). *Pay-as-you-go market attractiveness index 2019.* https://www.lightingglobal.org/wp-content/uploads/2019/06/PAYGo-MAI-2019-Report.pdf

Lofthouse, J., Simmons, R. T., & Yonk, R. M. (2015). Reliability of renewable energy: Geothermal. Logan. https://www.usu.edu/ipe/wp-content/uploads/2015/11/Reliability-Geothermal-Full-Report.pdf

Lusaka South Multi Facility Economic Zone. (2018). *Investors in the Zone, Lusaka South Multi Facility Economic Zone website.* Retrieved November 26, 2018, from http://www.lsmfez.co.zm/

Maamba Collieries. (2012). *Maamba collieries power generation project environmental and social impact assessment summary.*

Maamba Collieries. (2013). Maamba Collieries on schedule with power plant construction. *MCL News.* http://www.maambacoal.com/images/corporatemagazine/2013/Second2013.pdf

Maamba Collieries. (2015a, September). Grid upgrade works by ZESCO on course. *MCL News*. http://www.maambacoal.com/images/corporatemagazine/2015/Third2015.pdf

Maamba Collieries. (2015b). ZESCO starts power line upgrades to prepare for MCL power. *MCL News*. http://www.maambacoal.com/images/corporatemagazine/2015/Second2015.pdf

Mabumba, D. (2017). *Ministerial statement migration of electricity tariffs to cost-reflectivity*. National Assembly of Zambia. http://www.parliament.gov.zm/node/7020

MacroTrends. (2020). *Copper Prices—45 Year Historical Chart, MacroTrends.net*. Retrieved June 6, 2020, from https://www.macrotrends.net/1476/copper-prices-historical-chart-data

Mahoney, J. (2000). Path dependence in historical sociology. *Theory and Society, 29*(4), 507–548. http://www.jstor.org/stable/3108585

Mar, L. E. (2009). *Carbon impact of proposed hydroelectric dams in Chilean Patagonia*. Massachusetts Institute of Technology. https://dspace.mit.edu/handle/1721.1/53068

Marchese, A. J., & Zimmerle, D. (2018, July 2). The US natural gas industry is leaking way more methane than previously thought. Here's why that matters. *The Conversation*. https://theconversation.com/the-us-natural-gas-industry-is-leaking-way-more-methane-than-previously-thought-heres-why-that-matters-98918

Marconi, N., de Reis, C. F. B., & de Araújo, E. C. (2016). Manufacturing and economic development: The actuality of Kaldor's first and second laws. *Structural Change and Economic Dynamics, 37*, 75–89. https://doi.org/10.1016/j.strueco.2015.12.002

Mason, E. S., & Asher, R. E. (1973). *The World Bank since Bretton Woods*. The Brookings Institution.

Matanzima, J. (2021). Thayer Scudder's Four Stage Framework, water resources dispossession and appropriation: The Kariba case. *International Journal of Water Resources Development, 00*(00), 1–24. https://doi.org/10.1080/07900627.2020.1866505

McKinney, J., et al. (2020). *Sustainability of China's overseas special economic zones: Zambia, National Geospatial-Intelligence Agency*. Retrieved August 9, 2020, from https://www.tearline.mil/public_page/sustainability-of-chinas-overseas-special-economic-zones-zambia/

McKinsey. (2019). *Second-life EV batteries: The newest value pool in energy storage*. McKinsey & Company. https://www.mckinsey.com/industries/automotive-and-assembly/our-insights/second-life-ev-batteries-the-newest-value-pool-in-energy-storage

McMillan, M. S., & Rodrik, D. (2011). Globalization, structural change and productivity growth. https://doi.org/10.3386/w17143

McNamara, R. S. (1966, May 18). Security in the contemporary world. Address of the US Secretary of Defense to the American Society of Newspaper Editors.

Mfula, C. (2010a). First Quantum's Zambia mine hit by power outage. *Reuters.* https://www.reuters.com/article/update-2-zambia-mining-power-idAFLDE6180X820100209

Mfula, C. (2010b, August 21). Singapore firm to spend $108 mln on Zambia coal mine. *Reuters.* https://uk.reuters.com/article/mining-zambia/singapore-firm-to-spend-108-mln-on-zambia-coal-mine-idUKLDE67K09320100821

Mfula, C. (2016, May 10). Zambia to diversify generation mix as drought hits Hydropower. *Lusaka Times.* https://www.lusakatimes.com/2016/05/10/zambia-diversify-generation-mix-drought-hits-hydropower/

Ministry of Commerce Trade and Industry. (2014). *Manufacturing Sector Study Report: 2011–2012.*

Ministry of Energy. (2018). *Strategic plan 2018–2021.* https://www.moe.gov.zm/wp-content/uploads/2019/09/Ministry-of-Energy-Strategic-Plan-2018-2021.pdf

Ministry of Energy and Water Development. (2008). *National energy policy.* https://doi.org/10.1017/CBO9781107415324.004

Ministry of Energy and Water Development. (2011). *Power system development master plan for Zambia.* https://www.moe.gov.zm/download/other-documents/Power-Systems-Development-Master-Plan.pdf

Ministry of National Development Planning. (2017). *Seventh National Development Plan 2017–2021.* http://www.lse.ac.uk/GranthamInstitute/wp-content/uploads/2018/09/8721.pdf

Moloney, A. (2019, September 27). In Honduras, years of drought pressure farmers to leave land. *Reuters.* https://uk.reuters.com/article/us-climate-change-honduras-migrants-anal/in-honduras-years-of-drought-pressure-farmers-to-leave-land-idUSKBN1WC0M6

Morrell, D. (1987). *Indictment: Power and politics in the construction industry.* Faber and Faber Ltd..

Moyo, D. (2013). *Winner take all: China's race for resources and what it means for us.* Penguin.

Mulanda, S. M., & Punt, C. (2020). *Structural change and economic dynamics structural characteristics of Zambia's agricultural sector and the role for agricultural policy : Insights from SAM based modelling.* Stellenbosch.

Mwila, A., Sinyenga, G., Buumba, S., Muyangwa, R., Mukelabai, N., Sikwanda, C., Chimbaka, B., Banda, G., Nkowani, C., & Bwalya, B. (2017). *Impact of load shedding on small scale enterprises.* [online] Lusaka: Energy Regulation Board. Available at: [Accessed 22 June 2021].

Naqvi, S. N., & Ahmed, A. M. (1986). *Preliminary revised PIDE macro-econometric model of Pakistan's economy.* Institution of Development Economics.

National Breweries. (2016). *Annual Report 2015.*

National Breweries. (2018) *Annual Report 2017.*

National Grid. (2015). *Electricity capacity report.* https://doi.org/10.1016/j.matchemphys.2007.04.026

Ncube, M., & Lufumpa, C. L. (2017). *Infrastructure in Africa—Lessons for future development.* Policy Press.

Ncube, M., Lufumpa, C. L., & Kararach, G. (2017). Introduction: Infrastructure in African development. In M. Ncube & C. L. Lufumpa (Eds.), *Infrastructure in Africa—Lessons for future development.* Policy Press.

Nelson, R., & Winter, S. (1982). *An evolutionary theory of economic change.* Harvard University Press.

North, D. C. (1990). Institutions, institutional change and economic performance—Political economy of institutions and decisions. Kindle. University of Cambridge Press.

North, D. C., & Weingast, B. R. (1989). Constitutions and commitment: The evolution of institutions governing public choice in seventeenth-century England. *The Journal of Economic History, 49*(4), 803–832. https://doi.org/10.1017/S0022050700009451

Ocko, I. B., & Hamburg, S. P. (2019). Climate impacts of hydropower: Enormous differences among facilities and over time. *Environmental Science and Technology, 53*(23), 14070–14082. https://doi.org/10.1021/acs.est.9b05083

OECD. (2019). Aid (ODA) by sector and donor, OECD website.

Olivero, J., et al. (2017). Recent loss of closed forests is associated with Ebola virus disease outbreaks. *Nature Scientific Reports, 7*(1), 1–9. https://doi.org/10.1038/s41598-017-14727-9

Olivier, R. (1961a). Transcript of interview with Burke Knapp.

Olivier, R. (1961b). Transcript of interview with Richard H. Demuth. .

Olley, G. S., & Pakes, A. (1996). The dynamics of productivity in the telecommunications equipment industry. *Econometrica, 64*(6), 1263. https://doi.org/10.2307/2171831

Oseni, M. O., & Pollitt, M. G. (2015). A firm-level analysis of outage loss differentials and self-generation: Evidence from African business enterprises. *Energy Economics, 52*, 277–286. https://doi.org/10.1016/j.eneco.2015.11.008

Ovo Energy. (2018). *Nuclear power explained, Ovoenergy.com.* Retrieved January 1, 2021, from https://www.ovoenergy.com/guides/energy-sources/nuclear-energy.html#:~:text=The potential power locked inside,create aridiculous24%2C000%2C000kWh.

Park, S. F. (1968). *Basic data on the economy of Zambia.* https://play.google.com/books/reader?id=K_ZhUrU5lfoC&hl=en_GB&pg=GBS.PA1

Pasha, H. A., Ghaus, A., & Malik, S. (1989). The economic cost of power outages in the industrial sector of Pakistan. *Energy Economics, 11*(4), 301–318. https://doi.org/10.1016/0140-9883(89)90046-7

Patil, R. (2017, July 28). Nava Bharat Ventures up 2%, achieves provisional acceptance of 300 MW power plant. *Money Control*. https://www.moneycontrol.com/news/business/stocks/nava-bharat-ventures-up-2-achieves-provisional-acceptance-of-300-mw-power-plant-2338365.html

Phiri, C. (2019, March). No tariff hike yet, it's only a proposal, says minister. *Zambia Reports*.

Pierson, P., & Skocpol, T. (2002). Historical institutionalism in contemporary political science. In Katznelson, I. and Milner, H. (eds.), *Political Science—State of the Discipline*, pp. 693–721. American Political Science Association and WW Norton & Company.

Piketty, T. (2014). *Capital in the twenty-first century* (A. Goldhammer, Ed.). The Belknap Press of Harvard University Press.

Plott, C. R., & Zeiler, K. (2005). The willingness to pay-willingness to accept gap, the 'endowment effect,' subject misconceptions, and procedures for eliciting valuations. *The American Economic Review, 95*(3), 530–545.

Power Technology. (2013, October 6). AfDB approves $150m loan for 300 MW Maamba power project and coal mine in Zambia. *Power Technology*. Retrieved from https://www.power-technology.com/uncategorised/newsafdb-approves-150m-loan-for-300mw-maamba-power-project-and-coal-mine-in-zambia/

Pravalie, R., & Bandoc, G. (2018). Nuclear energy: Between global electricity demand, worldwide decarbonisation imperativeness, and planetary environmental implications. *Journal of Environmental Management, 209*, 81–92. https://doi.org/10.1016/j.jenvman.2017.12.043

Reinert, E. S. (2019). German economics as development economics: From the thirty years' war to World War II. In R. Kattel (Ed.), *The visionary realism of German economics: from the thirty years' war to the cold war* (pp. 15–36). Anthem Press. https://doi.org/10.2307/j.ctvcb5979

Reinikka, R., & Svensson, J. (2002). Coping with poor public capital. *Journal of Development Economics, 69*(1), 51–69. https://doi.org/10.1016/S0304-3878(02)00052-4

Resnick, D., & Thurlow, J. (2014). *The political economy of Zambia's recovery: Structural change without transformation?* http://ebrary.ifpri.org/utils/getfile/collection/p15738coll2/id/128012/filename/128223.pdf

Reuters. (2013, January 7). Zambia's Sata says high maize meal price could spark riots. *Reuters*. https://www.reuters.com/article/zambia-riots-idUSL5E9C75HI20130107

Richter, A. (2020, May 13). Developer Kalahari GeoEnergy Ltd. has secured a convertible loan of USD 3.2 m for a up to three slim wells to be drilled for the company's Bweengwa geothermal power project in Zambia, Africa. *Think GeoEnergy*. Retrieved from https://www.thinkgeoenergy.com/kalahari-geoenergy-secures-3-2m-loan-funding-for-geothermal-project-in-zambia/

Rodrik, D. (2016). Premature deindustrialization. *Journal of Economic Growth, 21*(1), 1–33. https://doi.org/10.1007/s10887-015-9122-3

Rojas, O., Li, Y., & Cumani, R. (2014). *Understanding the drought impact of El Niño on the global agricultural areas: An assessment using FAO's Agricultural Stress Index (ASI)*. http://www.fao.org/3/a-i4251e.pdf

Roosevelt, F. D. (1945, February). Message to congress on the Bretton woods agreements. The American Presidency Project, 12. https://www.presidency.ucsb.edu/node/210026

Rosenstein-Rodan, P. N. (1943). Problems of industrialisation of Eastern and South-Eastern Europe. *The Economic Journal, 53*(210/211), 202–211. http://www.jstor.org/stable/2226317

Rostow, W. W. (1960). *The stages of economic growth: A non-communist manifesto* (3rd ed.). Cambridge University Press.

Sabatier, P. A. (1998). The advocacy coalition framework: Revisions and relevance for Europe. *Journal of European Public Policy, 5*(1), 98–130. https://doi.org/10.1080/13501768880000051

SADC. (2008). *Integrated water resources management strategy and implementation plan for the Zambezi River Basin*. http://www.zambezicommission.org/sites/default/files/clusters_pdfs/Zambezi%20River_Basin_IWRM_Strategy_ZAMSTRAT.pdf.

Samuelson, P. A. (1962). Parable and realism in capital theory: The surrogate production function. *The Review of Economic Studies, 29*(3), 193–206. https://doi.org/10.2307/2295954

Sanders, D. et al. (2016). *An analysis of electricity system flexibility for Great Britain*. https://www.gov.uk/government/uploads/system/uploads/attachment_data/file/568982/An_analysis_of_electricity_flexibility_for_Great_Britain.pdf

Sandwell, P., et al. (2016). Analysis of energy access and impact of modern energy sources in unelectrified villages in Uttar Pradesh. *Energy for Sustainable Development. International Energy Initiative, 35*, 67–79. https://doi.org/10.1016/j.esd.2016.09.002

Sanghvi, A. P. (1982). Economic costs of electricity supply interruptions. US and foreign experience. *Energy Economics, 4*(3), 180–198. https://doi.org/10.1016/0140-9883(82)90017-2

Savin, I., Marson, M., & Sutormina, M. (2020). How different aid flows affect different trade flows: Evidence from Africa and its largest donors. *Structural Change and Economic Dynamics, Elsevier, 55*(C), 119–136.

Scherer, L., & Pfister, S. (2016). Hydropower's biogenic carbon footprint. *PLoS ONE, 11*(9), 1–12. https://doi.org/10.1371/journal.pone.0161947

Schlesinger, A. J. (1949). *The Vital Center: The politics of freedom*. Houghton Mifflin.

Scott, J. C. (1998). *Seeing like a state—How certain schemes to improve the human condition have failed*. Yale University Press.

Scudder, T. (1993). Pipe Dreams: Can the Zambezi River supply the region's water needs?. Cultural Survival Quarterly, [online] (June). Available at: [Accessed 23 June 2021].

Scudder, T., & Colson, E. (1982). From welfare to development: A conceptual framework for the analysis of dislocated people. In A. Hansen & A. Oliver-Smith (Eds.), *Involuntary migration and resettlement* (pp. 267–288). Westview Press.

Serpell, N. (2019). *The old drift* (1st ed.). Hogarth.

Sharma. (2010). *Note to users.* https://doi.org/10.1177/0010880481022002l4

Sharma, P. A. (2017). *Robert McNamara's other war—The World Bank and International Development.* University of Pennsylvania Press.

Sichone, Y. M., et al. (2016). Electricity load shedding. An econometric analysis of the productivity of firms in the manufacturing sector in Lusaka. *International Journal of Commerce and Management Research, 2*(12), 151–157.

Sikwanda, C. (2019). *Nurturing the manufacturing sector: Ensuring strong roots for Zambian growth.* International Growth Centre.

Simpasa, A. et al. (2013). *Capturing mineral revenues in Zambia: Past trends and future prospects capturing mineral revenues in Zambia: Past trends and future prospects.* https://www.un.org/en/land-natural-resources-conflict/pdfs/capturing-mineral-revenues-zambia.pdf

Sinclair-Desgagné, B., & Soubeyran, A. (2000). *A Theory of Routines as Mindsavers,* 52. Montréal. https://www.researchgate.net/publication/4816610_A_Theory_of_Routines_as_Mindsavers

Sinkamba, K. (2011, September 2). Zambia approves power purchase agreement between ZESCO and Maamba Collieries. *SteelGuru.* https://steelguru.com/steel/zambia-approves-power-purchase-agreement-between-zesco-and-maamba-collieries/223295

Solow, R. M. (1956). A contribution to the theory of economic growth. *The Quarterly Journal of Economics, 70*(1), 65–94. http://www.jstor.org/stable/1884513

Spalding-Fecher, R. (2012). *Powering Southern Africa—The future of hydro in the Zambezi River Basin, Climate & Development Knowledge Network.* Retrieved June 6, 2020, from https://cdkn.org/2012/02/powering-southern-africa-the-future-of-hydro-in-the-zambezi-river-basin/?loclang=en_gb

Spalding-Fecher, R., et al. (2016). The vulnerability of hydropower production in the Zambezi River Basin to the impacts of climate change and irrigation development. *Mitigation and Adaptation Strategies for Global Change, 21*(5), 721–742. https://doi.org/10.1007/s11027-014-9619-7

Staples, A. L. S. (2006). *The birth of development: How the World Bank, Food and Agriculture Organisation and World Health Organisation changed the world, 1945–1965.* The Kent State University Press.

Steel News. (2011). Konkola Copper Mines to invest USD 33 million for diesel generators. *Steel Guru*. https://steelguru.com/steel/konkola-copper-mines-to-invest-usd-33-million-for-diesel-generators/242497

Steinbuks, J., & Foster, V. (2010). When do firms generate? Evidence on in-house electricity supply in Africa. *Energy Economics, 32*(3), 505–514. https://doi.org/10.1016/j.eneco.2009.10.012

Stiglitz, J. E. (2017). Promoting Sustainable Industrial Policies. In *Industrialize Africa: Strategies, Policies, Institutions and Financing*, pp. 11–15 African Development Bank Group. https://www.afdb.org/en/news-and-events/industrialize-africa-strategies-policies-institutions-and-financing-17570/

Stinchcombe, A. L. (1968). *Constructing social theories*. University of Chicago Press.

Strydom, A., & Morar, G. (2018). *Rooftop solar: While municipalities focus on feed-in tariffs, a much larger problem persists, Sustainability Institute*. Retrieved March 17, 2020, from https://www.sustainabilityinstitute.net/si-news/5503-rooftop-solar-while-municipalities-focus-on-feed-in-tariffs-a-much-larger-problem-persists

Sutton, W. et al. (2019) *Striking a balance*. World Bank. https://doi.org/10.1177/1941738119889627

Syndigate. (2011, June 17). KCM deploys 3 diesel generators at Konkola Deep Mining Project. *MENA Report*. https://search-proquest-com.libproxy.ucl.ac.uk/docview/872316305?rfr_id=info%3Axri%2Fsid%3Aprimo

Szirmai, A. (2012). Industrialisation as an engine of growth in developing countries, 1950–2005. *Structural Change and Economic Dynamics, 23*(4), 406–420. https://doi.org/10.1016/j.strueco.2011.01.005

Szirmai, A., & Verspagen, B. (2015). Manufacturing and economic growth in developing countries, 1950–2005. *Structural Change and Economic Dynamics, 34*, 46–59. https://doi.org/10.1016/j.strueco.2015.06.002

Takouleu, J. M. (2020, May 25). REPP finances drilling of 3 geothermal wells on Bweengwa site. *Afrik21*. https://www.afrik21.africa/en/zambia-repp-finances-drilling-of-3-geothermal-wells-on-bweengwa-site/

Teljeur, E., Chetty, M., & Hendriksz, M. (2017). Africa's prospects for infrastructure development and regional integration: Energy sector. In M. Ncube & C. L. Lufumpa (Eds.), *Infrastructure in Africa—Lessons for Future Development* (1st ed., pp. 185–256). Policy Press.

Tembo, B. (2018). *Strategic investment decisions in Zambia's mining sector under a constrained energy system*. University College London. https://discovery.ucl.ac.uk/id/eprint/10048151/

Tembo, B., & Merven, B. (2013). Policy options for the sustainable development of Zambia's electricity sector. *Journal of Energy in Southern Africa, 24*(2), 16–27. https://doi.org/10.17159/2413-3051/2013/v24i2a3126

Thaler, R. (1980). Toward a positive theory of consumer choice. *Journal of Economic Behavior and Organization, 1*, 39–60. https://doi.org/10.1016/0167-2681(80)90051-7

Thurlow, J., & Wobst, P. (2007). The role of agriculture in pro-poor growth: Lessons from Zambia. In M. Grimm, S. Klasen, & A. McKay (Eds.), *Determinants of pro-poor growth—Analytical issues and findings from country cases* (pp. 219–241). Palgrave Macmillan.

Thurlow, J., Zhu, T., & Diao, X. (2012). Current climate variability and future climate change: Estimated growth and poverty impacts for Zambia. *Review of Development Economics, 16*(3), 394–411. https://doi.org/10.1111/j.1467-9361.2012.00670.x

Tobey, R. C. (1996). *Technology as freedom: The New Deal and the electrical modernization of the American home*. University of California Press.

Transparency International. (2020). *Corruption Perceptions Index 2019, Corruption Perceptions Index 2019.* Retrieved March 2, 2020, from https://www.transparency.org/country/ZMB

Trimble, C. et al. (2016). *Financial viability of electricity sectors in sub-Saharan Africa—Quasi-fiscal deficits and hidden costs, energy and extractives global practice group,* 7788. https://openknowledge.worldbank.org/handle/10986/24869

Tripsas, M., & Gavetti, G. (2000). Capabilities, cognition, and inertia: Evidence from digital imaging. *Strategic Management Journal, 21*(Special), 1147–1161. https://doi.org/10.1002/1097-0266(200010/11)21:10/11<1147::aid-smj128>3.0.co;2-r

Tushman, M. L., & Anderson, P. (1986). Technological discontinuities and organizational environments. *Administrative Science Quarterly, 31*(3), 439–465. https://doi.org/10.2307/2392832

U.S. Census Bureau. (1999). Table No. 1430. Employment Status of the Civilian Population: 1929 to 1998. In *Statistical abstract of the United States.* U.S. Census Bureau. https://www.census.gov/prod/99pubs/99statab/sec31.pdf

U.S. Geological Survey. (2015). *Mineral commodity summaries 2015, Mineral Commodity Summaries.* Reston, VI. https://s3-us-west-2.amazonaws.com/prd-wret/assets/palladium/production/mineral-pubs/copper/mcs-2015-coppe.pdf

U.S. Geological Survey. (2016). *Mineral commodity summaries 2016, Mineral Commodity Summaries.* Reston, VI. https://s3-us-west-2.amazonaws.com/prd-wret/assets/palladium/production/mineral-pubs/copper/mcs-2016-coppe.pdf

U.S. Geological Survey. (2017). *Mineral commodity summaries 2017, Mineral Commodity Summaries.* Reston, VI. https://s3-us-west-2.amazonaws.com/prd-wret/assets/palladium/production/mineral-pubs/copper/mcs-2017-coppe.pdf

U.S. Geological Survey. (2018). *Mineral commodity summaries 2018, Mineral Commodity Summaries.* Reston, VI. https://s3-us-west-2.amazonaws.com/prd-wret/assets/palladium/production/mineral-pubs/copper/mcs-2018-coppe.pdf

U.S. Geological Survey. (2019). *Mineral Commodity Summaries 2019, Mineral Commodity Summaries.* Reston, VI. https://prd-wret.s3-us-west-2.amazonaws.com/assets/palladium/production/s3fs-public/atoms/files/mcs-2019-coppe.pdf

U.S. Geological Survey. (2020). *Mineral Commodity Summaries 2020.* Reston, VI. https://pubs.usgs.gov/periodicals/mcs2020/mcs2020-copper.pdf

UCLA Statistics. (2019). *F Distribution tables, statistical online computational resource.* Retrieved April 14, 2020, from http://www.socr.ucla.edu/Applets.dir/F_Table.html

UN. (2008). International standard industrial classification of all economic activities. New York. https://unstats.un.org/unsd/publication/seriesm/seriesm_4rev4e.pdf

United Nations. (1962). *International Trade Statistics 1900–1960, UN Trade Statistics.* https://unstats.un.org/unsd/trade/data/tables.asp

United Nations. (2017). *Household size and composition around the world, Economic & Social Affairs.* https://doi.org/10.3390/atmos6091362

US Bureau of Labor Statistics. (2018). Table 24. Historical Consumer Price Index for All Urban Consumers (CPI-U): U. S. city average, all items. US Bureau of Labor Statistics.

US EPA. (2020a). *Understanding Global Warming Potentials, US Environmental Protection Agency website.* Retrieved June 6, 2020, from https://www.epa.gov/ghgemissions/understanding-global-warming-potentials#:~:text=Methane (CH4) is estimated,uses a different value.).&text=Nitrous Oxide (N2O,for a 100-year timescale

US EPA. (2020b). *Why is acid rain harmful? US Environmental Protection Agency website.* Retrieved August 10, 2020, from https://www3.epa.gov/acidrain/education/site_students/whyharmful.html#:~:text=Acid Rain Harms Forests,trees to take up water

USAID, Tony Blair Institute for Global Change and Power Africa. (2019). *West Africa Power Trade Outlook.* https://institute.global/sites/default/files/inline-files/ENGLISHWestAfricaPowerTradeUKWHOLELOW_1.pdf

Vidal, J. (2016, May 30). El Niño is over—But it leaves nearly 100 million people short of food. *The Guardian.* https://www.theguardian.com/global-development/2016/may/30/el-nino-is-over-but-it-leaves-nearly-100-million-people-short-of-food

Vivid Economics and Oeko-Institut. (2020). *Closing the loop on energy access: Analytical study.* Vivid Economics.

Wang, G., et al. (2017). Continued increase of extreme El Niño frequency long after 1.5°C warming stabilization. *Nature Climate Change, 7*(8), 568–572. https://doi.org/10.1038/NCLIMATE3351

Waylen, P., Annear, C., & Bunting, E. (2019). Interannual hydroclimatic variability of the lake. *Water, 11*(9), 1–14. https://doi.org/10.3390/w11091801

WFP. (2016). *El Niño: Undermining resilience, world food programme*. https://documents.wfp.org/stellent/groups/public/documents/ena/wfp282004.pdf?_ga=2.257185602.813928114.1602933689-1385311309.1602933689

Whitworth, A. (2014). Energy policy. In C. Adam, P. Collier, & M. Gondwe (Eds.), *Zambia: Building prosperity from resource wealth*. Oxford Scholarship Online. https://doi.org/10.1093/acprof:oso/9780199660605.003.0011

Willing, N. (2020, March). Chinese 5G rollout lifts lithium battery demand. *Argus Media*. https://www.argusmedia.com/en/news/2088490-chinese-5g-rollout-lifts-lithium-battery-demand

Willis, K. G., & Garrod, G. D. (1997). Electricity supply reliability: Estimating the value of lost load. *Energy Policy, 25*(1), 97–103.

Wobst, P., & Thurlow, J. (2005). The road to pro-poor growth in Zambia: Past lessons and future challenges. In *Proceedings of the German Development Economics Conference, Kiel 2005 / Verein für Socialpolitik, Research Committee Development Economics, No. 37*. International Food Policy Research Institute. http://hdl.handle.net/10419/19828

World Bank. (1999). Project appraisal document on a proposed loan and GEF grant in the amount of IBRD loan US$ 30 million and GEF Grant SDR 7.2 million to Argentine Republic for a Renewable Energy in the Rural Market Project. http://documents.worldbank.org/curated/en/271841468768697115/pdf/multi-page.pdf

World Bank. (2016). World Bank's First Development Loans to Chile, 1948, (057), pp. 1–5. http://documents.worldbank.org/curated/en/308691468185347709/pdf/104690-WP-PUBLIC-2007-01-World-Banks-First-Development-Loans-to-Chile.pdf

World Bank. (2017a) *Combined project information documents—Electricity Services Access Project P162760*. PIDISDSA21985.

World Bank. (2017b). *Investor brief*. DC. http://pubdocs.worldbank.org/en/284471507050951684/investor-brief-worldbank-2017.pdf

World Bank. (2017c). *World development indicators—Zambia*. Data Bank. https://data.worldbank.org/country/zambia

World Bank. (2017d). *Zambia solar radiation measurement data, EnergyData. Info*. Retrieved November 20, 2018, from https://energydata.info/dataset/zambia-solar-radiation-measurement-data-2015-2017

World Bank. (2018a). *Data on infrastructure—World bank enterprise survey of business managers, enterprise surveys*. Retrieved November 28, 2018, from http://www.enterprisesurveys.org/data/exploretopics/infrastructure

World Bank. (2018b). World Bank projects and operations in Zambia. World Bank. https://doi.org/10.1007/s10461-016-1645-x
World Bank. (2018c). *World development indicators—Zambia*. Data Bank. https://data.worldbank.org/country/zambia
World Bank. (2018d). Zambia precipitation 1991–2015. Climate Change Knowledge Portal, World Bank Group. http://sdwebx.worldbank.org/climateportal/index.cfm?page=downscaled_data_download&menu=historical
World Bank. (2019a). *World development indicators: Zambia*. World Bank. https://data.worldbank.org/country/zambia
World Bank. (2019b). *World development indicators—Chile*. https://data.worldbank.org/country/chile
World Bank. (2020a). *Electricity production from hydroelectric sources (% of total)*. https://data.worldbank.org/indicator/EG.ELC.HYRO.ZS?view=chart
World Bank (2020b) *GDP (constant LCU)*. https://data.worldbank.org/indicator/NY.GDP.MKTP.KN
World Bank. (2020c). *Metadata glossary, databank*. Retrieved October 24, 2020, from https://databank.worldbank.org/metadataglossary/wdi-database-archives-(beta)/series/NV.IND.TOTL.CD#:~:text=Longdefinition,electricity%2Cwater%2Candgas
World Bank. (2020d). *Projects in Bolivia, World Bank*. Retrieved October 17, 2020, from https://projects.worldbank.org/en/projects-operations/projects-list?lang=en&searchTerm=&countrycode_exact=BO
World Bank. (2020e). *Projects in Brazil, World Bank*. Retrieved October 17, 2020, from https://projects.worldbank.org/en/projects-operations/projects-list?lang=en&searchTerm=&countrycode_exact=BR
World Bank. (2020f). *Projects in Colombia, World Bank*. Retrieved October 17, 2020, from https://projects.worldbank.org/en/projects-operations/projects-list?lang=en&searchTerm=&countrycode_exact=CO
World Bank. (2020g). *Projects in Costa Rica, World Bank*. Retrieved October 17, 2020, from https://projects.worldbank.org/en/projects-operations/projects-list?lang=en&searchTerm=&countrycode_exact=CR
World Bank. (2020h). *Projects in Ecuador, World Bank*. Retrieved October 17, 2020, from https://projects.worldbank.org/en/projects-operations/projects-list?lang=en&searchTerm=&countrycode_exact=EC
World Bank. (2020i). *Projects in Ethiopia, World Bank*. Retrieved October 17, 2020, from https://projects.worldbank.org/en/projects-operations/projects-list?lang=en&searchTerm=&countrycode_exact=ET
World Bank. (2020j). *Projects in Ghana, World Bank*. Retrieved October 18, 2020, from https://projects.worldbank.org/en/projects-operations/projects-list?lang=en&searchTerm=&countrycode_exact=GH

World Bank. (2020k). *Projects in Guatemala, World Bank.* Retrieved October 18, 2020, from https://projects.worldbank.org/en/projects-operations/projects-list?lang=en&searchTerm=&countrycode_exact=GT

World Bank. (2020l). *Projects in Honduras, World Bank.* Retrieved October 18, 2020, from https://projects.worldbank.org/en/projects-operations/projects-list?lang=en&searchTerm=&countrycode_exact=HN

World Bank. (2020m). *Projects in Panama, World Bank.* Retrieved October 18, 2020, from https://projects.worldbank.org/en/projects-operations/projects-list?lang=en&searchTerm=&countrycode_exact=PA

World Bank. (2020n). *Projects in Peru, World Bank.* Retrieved October 18, 2020, from https://projects.worldbank.org/en/projects-operations/projects-list?lang=en&searchTerm=&countrycode_exact=PE

World Bank. (2020o). *Projects in Sudan, World Bank projects & operations.* Retrieved October 18, 2020, from https://projects.worldbank.org/en/projects-operations/projects-list?lang=en&searchTerm=&countrycode_exact=SD

World Bank. (2020p). *Projects in Turkey, World Bank projects & operations.* Retrieved October 18, 2020, from https://projects.worldbank.org/en/projects-operations/projects-list?lang=en&searchTerm=&countrycode_exact=TR

World Bank. (2020q). *Projects in Uruguay, World Bank projects & operations.* Retrieved October 18, 2020, from https://projects.worldbank.org/en/projects-operations/projects-list?lang=en&&searchTerm=&countrycode_exact=UY

World Bank. (2020r). *World Bank list of all projects.* https://projects.worldbank.org/en/country?lang=en&page

World Bank. (2020s). *World hydroelectric data.* World Bank. https://data.worldbank.org/

World Bank. (2020t). *Zambia Trade Summary 2018 Data, World Integrated Trade Solution.* Retrieved April 11, 2020, from https://wits.worldbank.org/CountryProfile/en/Country/ZMB/Year/LTST/Summary

World Bank. (2020u). *Zambia world development indicators.*

World Bank and International Finance Corporation. (2014). *Zambia country profile 2013.* World Bank.

Yamba, F. D., Walimwipi, H., & Mzezewa, C. (2011). Climate change / variability implications on hydroelectricity generation in the Zambezi River Basin. *Mitigation and Adaptation Strategies for Global Change, 16*(6), 617–628. https://doi.org/10.1007/s11027-011-9283-0

Zachary, S. (2016). *Least worst regret analysis for decision making under uncertainty, with applications to future energy scenarios.* Edinburgh. http://arxiv.org/abs/1608.00891

Zambeef Products. (2016). *Annual Report 2015.*

Zambeef Products. (2017). *Annual Report 2016.*

Zambeef Products. (2018). *Annual Report 2017.*

Zambia Bata. (2016). *Annual financial statements for the year ended 31 December 2015.*

Zambia Bata. (2017). *Annual Report 2016.*

Zambia Bata. (2018). *Abridged Report 2017.*

Zambia Chamber of Mines. (2016). *A concentrated mining sector, Mining for Zambia.* Retrieved October 1, 2019, from https://miningforzambia.com/a-concentrated-mining-sector/

Zambia Development Agency. (2019). *MSME Application Form.* Zambia Development Agency. http://www.zda.org.zm/?q=content/msme-application-form

Zambia Meteorological Department. (2018). *Raw precipitation data of 39 sites in Zambia, 1991–2018.*

Zambia Sugar. (2016). *2015 Annual Report.*

Zambia Sugar. (2017). *2016 Annual Report.* https://africanfinancials.com/document/zm-zmsg-2016-ar-00/

Zambia Sugar. (2018). *Annual Report for the period ended 31 August 2017.*

ZAMEFA. (2016a). Directors' Report and consolidated financial statements for the 9 month period ended 30 September 2016.

ZAMEFA. (2016b). Directors' Report and financial statements For the year ended 31 December 2015.

ZAMEFA. (2018). Directors' Report and consolidated financial statements for the year ended 30 September 2017.

ZCCM Investment Holdings Plc. (2019). *Mopani copper mines PLC, ZCCM-IH website.* Retrieved October 2, 2019, from https://zccm-ih.com.zm/copper-cobalt-gold/mopani-copper-mines/

Zeng, D. Z. (2016). *Multi-facility economic zones in Zambia: Progress, Challenges and possible interventions.* http://documents1.worldbank.org/curated/en/720981495115586647/pdf/115143-WP-PUBLIC-Feb-2016-GTCCS-ZambiaMFEZ.pdf

Zesco. (2019). *Countrywide load-shedding schedule October 2019 revised.* http://www.zesco.co.zm/customerCare/loadSheddingSchedule

Zesco Ltd. (2017a). *2016 Integrated Report.*

Zesco Ltd. (2017b). *Current Zesco tariffs, Zesco website.* Retrieved November 2018, from http://www.zesco.co.zm/customerCare/tariffs

Zesco Ltd. (2017c). *Revision of electricity tariffs in 2017.*

Zesco Ltd. (2017d). *Zesco's planned additional energy generation.* Zesco Ltd..

Zesco Ltd. (2017e). *Zesco load shedding schedules, Zesco website.* Retrieved November 1, 2018, from http://www.zesco.co.zm/customerCare/loadSheddingSchedule

Zesco Ltd. (2017f). *Zesco sales consumption by sector by month, 2010–2016*. Zesco Ltd.

Zesco Ltd. (2018). 2017 Integrated report.

ZETDC. (2014). *Electricity tariffs, ZETDC website*. Zimbabwe Electricity Transmission & Distribution Company.

ZICTA and CSO. (2018). *2018 National survey on access and usage of information and communication technologies by households and individuals—A demand side assessment of access and usage of ICTs in Zambia*. https://www.zicta.zm/Downloads/publications/2018ICTSurvey_PreliminaryReport.pdf

ZNBC. (2019). *Nurturing the manufacturing sector: Ensuring strong roots for Zambian growth*. ZNBC. https://www.youtube.com/watch?v=4VTlhgi713w&t=11s

Index[1]

[1] Note: Page numbers followed by 'n' refer to notes.

I. Ahmed, *The Political Economy of Hydropower Dependant Nations*,
https://doi.org/10.1007/978-3-030-71266-2

The manufacturer's authorised representative in the EU is Springer Nature Customer Service Centre GmbH, Europaplatz 3, 69115 Heidelberg, Germany. If you have any concerns regarding our products, please contact ProductSafety@springernature.com

Printed and bound by CPI Group (UK) Ltd, Croydon, CR0 4YY
30/09/2026
02225689-0002